Intelligent Seam Tracking for Robotic Welding

Nitin Nayak and Asok Ray

Intelligent Seam Tracking for Robotic Welding

With 75 Figures

Springer-Verlag
London Berlin Heidelberg New York
Paris Tokyo Hong Kong
Barcelona Budapest

Nitin R. Nayak, PhD, MS, BS
IBM Corporation,
Austin,
Texas, USA

Asok Ray, PhD, PE
Pennsylvania State University,
PA 16802, USA

ISBN 3-540-19826-1 Springer-Verlag Berlin Heidelberg New York
ISBN 0-387-19826-1 Springer-Verlag New York Berlin Heidelberg

British Library Cataloguing in Publication Data
A catalogue record for this book is available from the British Library

Library of Congress Cataloging-in-Publication Data
A catalog record for this book is available from the Library of Congress

© Springer-Verlag London Limited 1993
Printed in Germany

Typesetting: Camera ready by authors
69/3830—543210 Printed on acid-free paper

To our parents

SERIES EDITORS' FOREWORD

The series *Advances in Industrial Control* aims to report and encourage technology transfer in control engineering. The rapid development of control technology impacts all areas of the control discipline. New theory, new controllers, actuators, sensors, new industrial processes, computing methods, new applications, new philosophies, . . ., new challenges. Much of this development work resides in industrial reports, feasibility study papers and the reports of advanced collaborative projects. The series offers an opportunity for researchers to present an extended exposition of such new work in all aspects of industrial control for wider and rapid dissemination.

In the field of robotics, the classic repetitive manufacturing tasks which have been automated by robots are paint spraying and spot/seam welding for the car industry. This monograph reports advances in robotic science for the operation of seam welding. It presents a systematic treatment of the prevailing industrial technology and a new state of the art intelligent robotic seam welding prototype system on which the authors, Dr Nayak and Professor Ray, collaborated. The authors have made a determined effort to set their work in the context of international robotic seam welding research and conclude by reviewing seven other international prototype systems. The mix of specific research issues and the review of broader research activities reported makes this a particularly welcome contribution to the Series.

October, 1992   M. J. Grimble and M. A. Johnson,
Industrial Control Centre,
Glasgow, G1 1QE,
Scotland, U.K.

PREFACE

Robotic welding is critical to welding automation in many industries. "Blind" robotic welding systems, however, cannot adapt to changes in the joint geometry which may occur due to a variety of reasons. For example, in systems following pretaught weld paths, in-process thermal distortion of the part during welding, part fixturing errors, and out-of-tolerance parts will shift the original weld path, leading to poor quality welds. Essential to accurate seam tracking is some form of joint sensing to adjust the welding torch position in realtime as it moves along the seam. Realtime seam tracking is attractive from the perspective of improving the weld quality and also reducing the process cycle time. In this monograph, we have addressed the technological aspects of adaptive, realtime, intelligent seam tracking for automation of the welding process in the context of three-dimensional (3D) seams.

The work reported in this monograph builds upon the research conducted during the course of project ARTIST (acronym for Adaptive, RealTime, Intelligent Seam Tracker) at the Applied Research Laboratory of the Pennsylvania State University, USA. The research project ARTIST was sponsored by the BMY Corporation of York, Pennsylvania, USA, and the Commonwealth of Pennsylvania to address requirements of welding steel and aluminium plates and castings used in the manufacture of heavy duty battlefield vehicles. A prototype version of ARTIST was designed and developed for tracking and welding planar seams. Since ARTIST was expected to encounter mostly slip joints during welding, the algorithms for seam tracking are predominantly based on the analysis of vee-grooves. The objective was to demonstrate the proof of concept, develop a prototype of a seam tracking system, and integrate it with the welding equipment for realtime operation.

The ARTIST system comprised of a six degree-of-freedom robot (PUMA 560 robot manipulator with Unimate controller), a laser profiling gage (Chesapeake Laser System), a PC-AT microcomputer serving as the supervisory controller, and welding equipment (Miller Electric Delta-weld 450 welding controller and a Miller Electric S54A wire feeder). At the end of the one year development period, the system was capable of tracking planar seams. The promising results motivated us to extend the scope of this system to tracking general 3D vee-jointed seams. Based on the data collected from real samples, we developed and tested algorithms for interpreting joint features in range images under conditions of variable position and orientation relationship between the sensor and the 3D seam. The analysis of seam tracking error is based on our experience with the operation of the ranging sensor and real joint geometries.

This monograph covers up-to-date and relevant work in the area of intelligent seam tracking. In contrast to many seam tracking systems that have been developed in the past for operation within well-defined working environments, this monograph primarily addresses the tracking of seams in unstructured environments. Essential to tracking seams within such an environment is some form of joint sensing. Chapter 2 provides an overview of the various sensing techniques while Chapter 3 covers the basic principles of processing intensity and range images for extracting and interpreting joint features, and the development of 3D seam environment models. Chapters 4 and 5 discuss the various coordinate frames and robot motion control issues related to seam tracking. Implementation details regarding development of a seam tracking system based on off-the-shelf components are presented in Chapter 6 and the various tracking errors are analyzed in Chapter 7. Finally an overview of the approaches used in existing seam tracking systems is presented in Chapter 8, and possible directions for future intelligent, realtime seam tracking are discussed in Chapter 9.

This monograph is directed towards readers who are interested in developing intelligent robotic applications. Although this work is presented in the context of seam tracking, the issues related to systems integration are general in nature and apply to other robotic applications as well.

November, 1992

Nitin R. Nayak
IBM Corporation
11400 Burnet Road
Austin, TX 78758, USA

Asok Ray
Mechanical Engineering Department
The Pennsylvania State University
University Park, PA 16802, USA

Acknowledgements

We are grateful to Mr. Henry Watson, Head of the Manufacturing Science and Technology department at the Applied Research Laboratory of the Pennsylvania State University, and the BMY Corporation, York, Pennsylvania, for their support during the initial phase of the ARTIST project. The hard work put in by Andy Vavreck to get this project going also deserves a lot of praise. Thanks are also due to Nancy Hahn and Prof. Gang Yu for providing the use of word processing facilities so crucial to the completion of this monograph. Finally, the support provided by the management at IBM Corporation, especially Ann Feicht and Peter Stoll, during the preparation of this monograph is highly appreciated.

CONTENTS

CHAPTER 1

INTRODUCTION

In the present day global marketplace, manufacturing organizations are facing national as well as international competition, forcing them to further improve their performance. To this effect, the concepts of *computer-integrated manufacturing* (CIM) (O'Rourke 1987) have been introduced in various production environments with the intent of:

- Improving human productivity at all levels;
- Improving product quality;
- Improving capital resource productivity;
- Providing rapid response to the market demands.

The CIM strategy is to integrate the information bases of the various units of automation (e.g., design, engineering, manufacturing, and office administration including, accounting, marketing, and sales) within the traditional framework of manufacturing. In this respect, CIM can be viewed as a closed loop control system where a typical input is the order for a product and the corresponding output is the delivery of the finished product.

Automation of the physical production processes on the shopfloor is a key component of the CIM strategy for improving productivity. In this context, robots have played an important role in the automation of various operations. Robots have been successful in automating simple and repetitive operations while simultaneously enhancing the quality of manufactured products in many production areas. The use of robots is also highly desirable in hazardous manufacturing operations such as spray painting, welding, etc., which pose known health risks to human operators. Although attempts have been made to precisely define the functions and characteristics of a robot, it can be generally defined as any automated machine that operates with the aid of computer-based sensors, controllers and actuators.

Before proceeding into the details of using robots for welding applications, an overview is provided of the various production processes where robotic solutions have been successful in the industry (Chang et al., 1991).

1.1 INDUSTRIAL ROBOTIC APPLICATIONS

Spray Painting. Spray painting is the earliest and one of the most widely used applications of robots in industry as this task is hazardous to human health and safety and does not require much intelligence. In addition, a robot uses less paint and produces coatings which are much more uniform than is usually possible manually. Several robotic spray painters can be programmed to simultaneously execute a given task with each robot following a continuous path along a straight or curved line. A spray painting robot normally does not require the use of external sensors. However, the part to be painted must be accurately presented to the robot manipulator.

Grinding and Deburring. Grinding is necessary to produce a smooth surface for a good appearance or for producing parts within required tolerance after a rough operation such as arc welding. This job is well suited for a robot which can perform the assigned task sequentially with its previous operation, say, arc welding by simply replacing the welding torch with a rotary grinder. To provide a finished surface on metal castings and for removing any undesirable high spots, the robot is taught to move along a continuous path corresponding to the correct shape of the casting. Robots have also been successfully used for deburring operations wherein the unwanted material around the back side of a drilled hole is cleanly removed to leave a smooth surface. In general, for grinding operations, uncertainties exist in the dimensions of the workpiece. This necessitates the use of sensory information to accurately assess the contour of the part. For example, in smoothing an arc welding bead, simple tactile sensors provide information about the surface contour to the robotic grinders.

In addition to rotary grinding and deburring applications, robots are also used for routing, drilling, honing, polishing, tapping, nut running, and screw driving. Although preprogramming of tool points or paths is sufficient in many cases, exact placement of drilled holes, such as those on aircraft structures, requires a template and a compliant-wrist robot.

Parts Handling, Transfer, Sorting, and Inspection. Moving parts from one location to another or picking parts from conveyors and arranging them on pallets is one of the most common applications of robots. Other parts handling operations include feeding unfinished parts into machines such as a lathe, punch press, or a furnace. Such simple and repetitive operations are well-suited for robots although they could be dangerous for human workers.

Pick-and-place robots in sophisticated workcells can sort individual parts from batches of unsorted parts when, for either cost reduction reasons or because of tolerance variations in the manufacturing process, dissimilar parts are grouped together. Robots, equipped with appropriate grippers, acquire the parts and bring them to the workstations where a gaging device informs the robot controller of the type and dimensions of the parts in order to place them in the correct bin. In some applications, vision systems are used to determine the type and orientation of the part on a conveyor. The vision systems are located upstream of the robot and the part information is passed to the robot controller after inspection. This allows the manipulator to move to the correct location with appropriate gripper orientation and pick up the part from the moving conveyor.

Vision sensors mounted on robot endeffectors have been used to inspect finished parts or subassemblies in order to increase product quality. Examples are inspection of parts of automobile bodies and printed circuit boards used in electronic devices.

Assembly Operations. Assembling a group of parts into a subassembly or a finished product for bulk production is an extremely repetitive and mundane job for human workers. Servo-controlled robotic assemblers can execute this task using hand-eye coordination techniques and tactile sensing. There have been many applications of simple tasks in electronic industry where robotic assemblers are routinely used. However, as the complexity of assembly operations increases, sophisticated robots with external sensors (vision or tactile) and compliant wrists are necessary.

1.2 AN OVERVIEW OF ROBOTIC WELDING

As much as 40% of all industrial robots are being used for welding tasks (Ross 1984). Robotic welding is being initiated to satisfy a perceived need for high-quality welds in shorter cycle times. Simultaneously, the workforce can be shifted from welding to other production areas with higher potential productivity and better environmental quality.

Manual welding must be limited to shorter periods of time because of the required setup time, operator discomfort, and safety considerations. Consequently, manual arc welding is less than 30% of the total operator working time (Boillot 1985). In contrast, the welding time fraction is above 85% for a welding robot, yielding an increase in productivity by a

3

factor of 2.8, assuming that the same welding speed is followed in both the cases. As a result, robotic welding has become highly pervasive in the automobile industry.

Welding operations that can be effectively carried out by an industrial robot can be classified into two categories: (i) spot welding; and (ii) arc welding. In spot welding, the robot is first taught a sequence of distinct locations which are stored in the robot's memory. The robot sequentially positions the spot welding gun at these locations during the actual production cycle. Because of the irregularity of the parts to be welded, a (three-dimensional) wrist is often necessary for dexterous positioning of the spot welding gun. The use of heavy welding tools and the reasonably long reach required of the robot manipulator implies that the servomotors for joint movements should be sufficiently strong to avoid undesirable vibrations. However, since the robot activities are pre-taught, no sensory information is needed for feedback control.

The arc welding environment, on the other hand, may require the movement of the torch along irregular seams or the filling of wide joints. A continuous-path servo-controlled robot is often designed for a specific type of welding application. For robotic welding along preprogrammed paths, it is necessary that the parts to be welded be accurately positioned and correctly held in place in order to teach complex three-dimensional paths. Additionally, during welding the parts should be presented to the robotic welder in precisely the same position and orientation. If these stringent part positioning and path programming conditions can be met, then no position sensing is necessary.

Many robotic welding systems, however, do not adapt to realtime changes in joint geometry and therefore, have only limited success in many welding applications. Thermal distortion from the intense heat of the welding arc can cause these changes in the joint geometry. Such variations are also caused due to fixturing errors or improper preparation of the weld joints. Hence, to produce high quality welds through mechanization, strict tolerances are necessary in both joint preparation and fixturing the weldpieces. A solution to this problem requires sensing the joint geometry to properly position the welding torch along the seam in realtime. Techniques for joint sensing have been based on mechanical, electrical, magnetic, and optical sensors, with each method having specific advantages and disadvantages in a given production situation (Brown 1975; Richardson 1982). However, the two most commonly used techniques are through-the-arc sensing (which uses the arc itself to guide the torch) and vision

4

sensing. An overview of seam tracking technology is presented in Chapter 2.

Since the early days of robotic welding, much progress has been made, especially in the area of seam tracking, i.e., moving the torch correctly along the seam.Seam tracking Early systems required the use of separate learning and welding steps. However, this two-pass system becomes very time-consuming when complex seams have to be manually taught, and so requires a large batch-size to justify its use. Furthermore, such a system cannot adjust to the variations in joint geometry that result primarily from thermal distortion during arc welding.

Efficiency of robotic welding can be increased if both, sensing and welding of the seam can be carried out in the same pass. This is clearly an improvement over two-pass welding (popularly known as *first generation* welding systems), where the first pass is dedicated to learning the seam geometry followed by actual tracking and welding in the second pass. The *second generation* welding systems, on the other hand, track the seam in realtime and are characterized by the absence of a separate learning phase i.e., the welding system simultaneously learns and tracks the seam.

The second generation systems, however, are capable of operating only in well-defined welding environments. Some systems accomplish realtime seam tracking by exploiting a special feature of the robot or by dedicating their application to a particular type of seam. In one application, the seam is constrained to lie on a cylindrical surface (Bollinger, et al. 1981). Yet another scheme for controlling the welding torch's position and velocity is applicable only for two-axis control, thereby constraining the seam to lie in a plane (Tomizuka 1980).

The advances in the realtime seam tracking techniques of the second generation are characterized by the following features (Niepold 1979; Linden and Lindskog 1980; Vanderbrug, et al. 1979):

- Usage of vision sensing,
- Techniques to reduce the adverse effects of optical noise from the welding arc,
- Usage of pattern recognition algorithms for extracting the seam's features from its images.

The next generation of welding systems is required not only to operate in realtime but also to learn rapidly changing seam geometries while operating within unstructured environments. In contrast to well-defined working environments, wherein the seam-related information

5

associated with each image is known *a priori*, the unstructured environment is characterized by the absence of this knowledge. The system has knowledge about all the possible scenarios it may encounter along the seam, but the actual sequence of their occurrence is unknown. These welding systems belong to the so called *third generation*, and use concepts from artificial intelligence (AI) to learn and model the global seam environment (Begin, et al. 1985) in order to control the torch-seam interaction.

1.3 OUTLINE OF THE MONOGRAPH

This monograph covers up-to-date and relevant work in the area of intelligent seam tracking, and specifically presents the development of a third generation seam tracking system for robotic welding, called the Adaptive, Realtime, Intelligent Seam Tracker . ARTIST is essentially a single-pass system, characterized by the absence of an additional teaching phase. It essentially comprises of a welding torch and a laser range sensor mounted on the endeffector of a six-axis robot. A seam in space can be traced by a five degree-of-freedom (DOF) robot manipulator. However, realtime seam tracking requires that both the sensor and the torch trace the seam, thus requiring a six DOF robot. As illustrated in Figure 1.1, the spatial relationship between the welding torch and the ranging sensor permits scanning the region immediately ahead of the torch. The bracket supporting the torch-sensor assembly is mounted on the endeffector and allows the ranging sensor to rotate ±90 degrees about the torch axis for proper orientation relative to the seam.

The following chapters discuss issues related to the various aspects of realtime intelligent seam tracking in an unstructured environment. Chapter 2 provides an overview of the various applicable sensing techniques while Chapter 3 covers the basics of processing intensity and range images for extracting and interpreting joint features, and the development of 3D seam environment models. Chapters 4 and 5 discuss the various coordinate frames and robot motion control issues, respectively, as related to seam tracking. A discussion on the implementation details for developing a seam tracking system based on off-the-shelf components is discussed in Chapter 6, and the analysis of various tracking errors is presented in Chapter 7. Existing seam tracking systems developed across the world are reviewed in Chapter 8 for the benefit of the reader and possible future directions for intelligent, realtime seam tracking are presented in Chapter 9. The reader may note that the issues related to

6

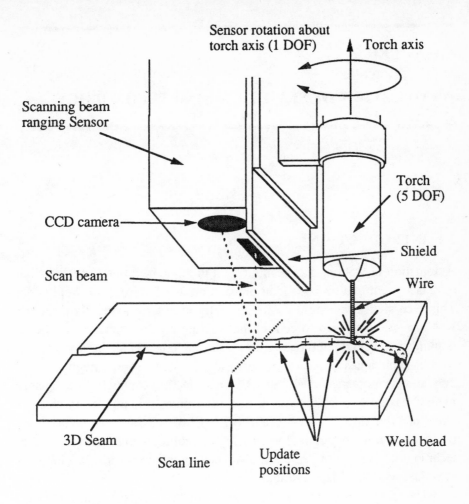

Figure 1.1 Robot endeffector assembly consisting of the look-ahead range sensor and the torch-tip.

welding process control are outside the scope of this monograph which is primarily dedicated to the problem of seam tracking in third generation robotic welding systems.

CHAPTER 2

AN OVERVIEW OF SEAM TRACKING TECHNOLOGY

2.1 INTRODUCTION

Automation of welding has been difficult to implement due to the extreme tolerances required on both joint preparation and fixturing the weld-parts. This necessitates sensing the joint geometry in order to correctly place the welding torch and also to control the welding process parameters. Seam tracking for welding automation has been implemented using a variety of sensing techniques based on mechanical, electrical, sonic, magnetic, and optical sensors (Jones et al. 1981; Justice 1983). Although each sensing method has its distinct advantages and disadvantages in a given production situation, the through-the-arc and 3-D vision sensing techniques have become the most popular approaches. In addition to the various sensing techniques, research efforts have also been directed towards development of robust seam tracking algorithms.

This chapter presents an overview of two sensing techniques, namely, through-the-arc sensing and vision sensing. Although ultrasonic ranging for seam tracking has been shown feasible by some researchers (Estochen, Neuman, and Prinz 1984; Fenn and Stroud 1986; Umeagukwu et al. 1989), it suffers from serious limitations in welding environments and so requires further research and investigation. To provide more information on this sensing technique, we have presented in Chapter 8 the details of an ultrasonic ranging-based seam tracking system prototype. In addition to the two sensing techniques mentioned above, this chapter also covers the corresponding seam tracking algorithms encountered in the literature. An overview of the evolution of the various seam tracking strategies for vision-based systems is also presented. The contents of this chapter set the stage for further detailed discussion on intelligent seam tracking for robotic welding which is presented in the following chapters.

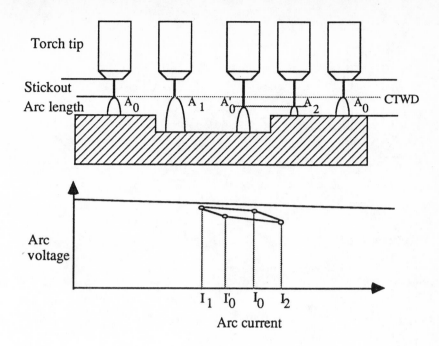

Figure 2.1. Arc Variations over idealized stepped plate.

2.2 THROUGH-THE-ARC SENSING

This sensing technique uses the functional relationship between the variations in the *Contact-Tip to Work Distance* (CTWD) and the electrical arc signals, i.e., arc current and arc voltage (Cook 1986). As depicted in Figure 2.1, when the torch moves from the steady state operating point **0** to point **1**, the CTWD increases. This results in an instantaneous increase of the arc length and the arc voltage, and a decrease in the arc current. Subsequently, the wire feed rate no longer equals the melting rate and the system acts to reestablish an equilibrium. Assuming a constant potential, self-regulating system, the operating point will return to within a few percent of the voltage and the current values that existed prior to the abrupt change in the CTWD. A similar but opposite change occurs for an abrupt decrease in the CTWD as depicted by movement from point **0'** to **2**.

The through-the-arc sensing is implemented for seam tracking by weaving (oscillating) the torch across the joint while also moving forward along a straight path parallel to the axis of the weld joint. The changes in

9

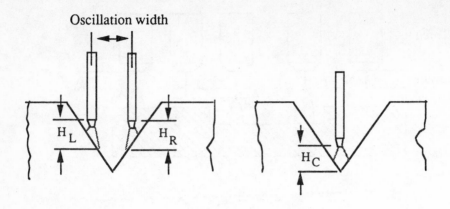

Figure 2.2 Sensing torch position via arc signal variations.

the joint geometry will be reflected in the corresponding changes in the CTWD, which in turn are related to changes in the arc voltage and arc current. This technique measures the joint geometry relative to the path of the electrode as shown in Figure 2.2.

The weaving action so necessary for the working of through-the-arc systems is actually helpful in providing an even heat distribution and hence a better weld quality. The through-the-arc sensing technique is especially useful in Submerged Arc Welding (SAW) since optical sensing is not supported by this process because the electrode, joint sides, arc, and the molten pool are hidden from direct viewing (Cook 1980). Generally, this technique has the advantage of directly sensing the local environment at the torchtip which allows automatic control of the bead placement, bead geometry, and fusion. Unfortunately, this also implies that the only available information about the seam is local and the equally important global knowledge (such as height mismatch, root gap, presence of tacks along the joint, etc.) is unavailable.

It should be noted that the through-the-arc sensing is not easily and economically implemented in all types of welding processes and weld joints. In particular, butt joints and shallow seams are difficult to sense using this technique. The operation of this system depends on measuring relatively small changes in the steady-state operating condition resulting from the change in CTWD. The rate at which the system reestablishes the new steady-state operating point is primarily a function of (1) the power source characteristics, (2) the arc characteristics, (3) the electrode extension, (4) the wire size, and (5) the properties of the wire material. Unfortunately, for most Gas-Metal Arc Welding (GMAW) and Flux-Core Arc Welding (FCAW), the desirable welding conditions make the time constant of the self-regulating process shorter than the torch's period of oscillations. Hence, recording the transient arc current and arc voltage requires high sampling frequencies and sophisticated control algorithms. This problem is not so severe in the SAW process because the large electrode size results in lower current density and therefore a longer time constant for the self-regulating process. For the ideal situation, the time constant of the self-regulating process should be comparable to the oscillation period of the torch to provide completely transient electrical signal variations across the joint.

2.2.1 Through-the-Arc Seam Tracking Control Algorithms

A simple control algorithm (Cook 1986) for reliably correcting the electrode's position relative to a vee-joint uses (1) the peak values of the current at each extreme of the torch oscillation, and (2) the value of the current at the center of the oscillation The difference between the peak current values at the left and right sidewalls determines the direction and the amount of lateral correction to be applied during the next oscillation cycle. A modification to the above-mentioned peak-current measurement approach is used to simultaneously center the electrode and also control the width. This involves moving the electrode toward a sidewall until the current reaches the preset value corresponding to the correct half-width of the joint. The electrode is then directed towards the other sidewall and the process is repeated. For tracking planar seams using this technique, the use of two electrodes is recommended. The use of three or more electrodes for tracking seams with three-dimensional variations has also been reported in the literature (Raina 1988).

2.2.2 Hardware Implementation

Through-the-arc sensing hardware is commercially available in three different configurations: general purpose nonrobotic systems, bolt-on robotic system, and integrated robotic system (Cook 1983).

The general purpose system consists of two cross slides and a controller unit. One cross slide allows cross-seam weaving and seam tracking while the other slide is used for proximity control. The cross-seam slide is operated by a microcomputer which controls the rate, width, centering, and left and right dwell. It communicates with the arc-signal processing microcomputer for sampling synchronization and centering correction. The signal processing microcomputer also controls the sampling of arc signals, digital filtering, and computing the cross-seam and vertical position error correction.

The bolt-on robotic system is similar to the general purpose system in that it has miniaturized slides attached directly to the robot arm. The advantage of this approach is that the miniaturized slides allow much faster oscillations of the cross-slides and hence faster travel speed. The disadvantage, however, is that the welding robot's work envelope is reduced since the robot arm is constrained to always position the cross-slide perpendicular to the torch-travel direction and the proximity control slide perpendicular to the weave plane.

In the integrated through-the-arc robotic system, the sensing and the torch position control is integrated within the robot controller. The integrated system provides the cross-seam weave pattern and the position corrections directly through the robot arm and the through-the-arc sensing system is invisible to the observer. The sensing subsystem microcomputer is responsible for processing arc signals and communicating the corrective actions to the robot controller. Cross-seam synchronization signals communicated by the robot controller allow the sensing subsystem to sample the arc signals at appropriate times. The corrective actions subsequently computed are presented at regular intervals to the robot controller for implementation. The integrated system has the advantage of programming the weave pattern in any position. This requires that the robot system be taught the start point, stop point and a third point to describe the weave plane.

2.3 3D VISION SENSING

This sensing technique has been by far the most studied and implemented in seam tracking systems. The advantages of this sensing technique include sensing the seam environment independent of the welding process and the possibility of gathering global information about the seam environment for intelligent seam tracking and overall welding process monitoring and control. 3D vision measurements can be classified as direct or indirect. Direct measurements result in range data of the surface points being measured while in the case of indirect measurements, the range data has to

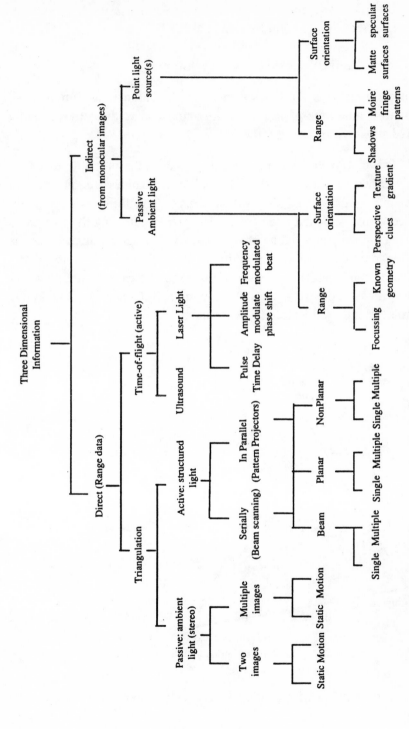

Figure 2.3 Techniques for measuring three-dimensional information (From Nitzan et al. 1987)

13

be inferred from monocular images. Figure 2.3 shows the taxonomy of these techniques in a tree form (Nitzan et al.1987).

2.3.1 Direct Range Measurement Techniques

Two different methods can be used to obtain direct range measurements—triangulation and time-of-flight. Both require that the surface being measured support lambertian reflection.

2.3.1.1 Triangulation: Triangulation is based on the principle that if at the two vertices of a triangle, the side and the two angles are known, then the distance from one of the two vertices to the third can be easily computed using elementary geometry. The plane containing the triangle is called the epipolar plane and its intersection with the camera image plane is called the epipolar line (refer Figure 2.4). 3D-vision systems using triangulation can be classified as stereo vision (if two cameras are placed at the two vertices of the triangle) and structured light (if one camera and one light projector are used instead of two cameras).

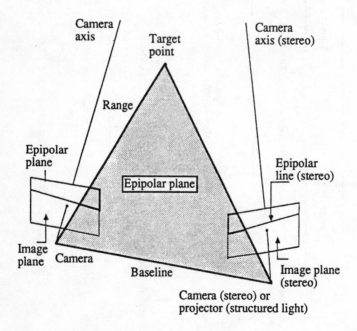

Figure 2.4 Principle of triangulation for range measurement (From Nitzan et al. 1987)

14

For arc welding applications, cameras which do not have complex multi-lens transfer components should be chosen to prevent unwanted internal reflections. Neutral density filters can be used to attenuate light across the entire spectrum thus allowing viewing of the arc. In the case of structured light systems, the selective spectrum filters centered around the laser beam wavelength are especially helpful since they only allow wavelengths around the reflected beam to pass even in the presence of a strong arc.

A serious problem with triangulation based systems is the shadow region where the surface being measured is not visible from both vertices of the triangle. The shadow region can be reduced by decreasing the baseline i.e., the distance between the two vertices of the triangle (refer Figure 2.4). However, this arrangement reduces the measurement accuracy. The measurement accuracy of the system can be increased by increasing the focal length of the cameras but this reduces the field of view. For any given application, a trade off has to be made regarding the measurement accuracy, the shadow region, and the field of view.

Stereo Vision: Stereo vision systems have two cameras and rely on ambient light to illuminate the surface being measured. In addition to the above problem of design tradeoffs with triangulation based systems, the stereo vision systems have difficulty in matching the corresponding surface points in the two camera images. This problem is even more acute when

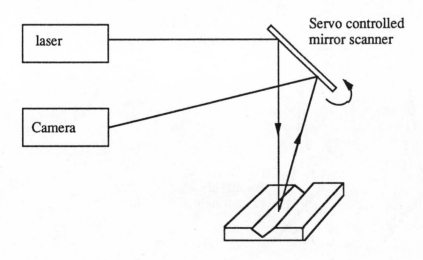

Figure 2.5 Principle of serial beam scanning ranging system.

the surface being measured has fewer facets and provides an even reflectance in the two images. For stereo vision to work well, it is desirable that the surface being measured have more intensity features such as edges, especially normal to the epipolar plane. The algorithms for matching corresponding points typically use both micro and macro constraints and have increased measurement time. This makes them infeasible for realtime applications.

Structured Light Systems: These systems replace one of the cameras in the stereo vision systems with a source of controlled illumination called structured light. The term structured light refers to a pattern of lines, triangles, or grids projected on the surface being measured. Commonly, the structured light may be projected serially on the surface by scanning a collimated laser beam or it can be a sheet of light generated by fanning a laser beam through a cylindrical lens or using a slit in a slide projector (refer Figures 2.5 and 2.6). Many commercial systems based on scanning a

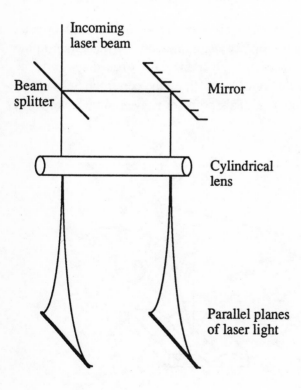

Figure 2.6 Structured light projection of dual sheets of light.

beam of light use mirrors attached to galvanometers, rotating cylinders, or use acousto-optic deflectors excited by radio frequency oscillations from a computer controlled oscillator. Serial systems are slow in measuring large areas while light sheet systems suffer from extensive matching to find the corresponding points in successive scans. However, for seam tracking purposes, the corresponding point problem is irrelevant if the joint features can be successfully identified within each image.

Structured light systems require that the surface being measured have Lambertian reflectance characteristics. As illustrated in Figure 2.7, Lambertian reflectance is a characteristic of diffuse surfaces although the amount of light absorbed may vary. These surfaces reflect light in all directions with maximum intensity along the normal to the illuminated surface while intensity along any other direction is given by equation (2.1).

$$I = I_0 \cos(\theta) \qquad (2.1)$$

where,

θ : is the angle of the reflected beam with the surface normal;
I_0 : is the intensity along the surface normal.

Most surfaces encountered in welding applications have mixed reflection with both, diffuse and specular reflection characteristics as depicted in Figure 2.8. Structured light systems are difficult to use with surfaces having high specular reflection content. This is due to the fact that the reflected light beam will not reach the receiving camera (resulting in

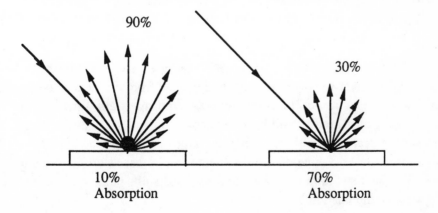

Figure 2.7 Diffuse reflection from partially absorbing Lambertian surfaces.

17

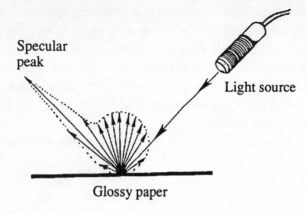

Specular peak

Light source

Glossy paper

Figure 2.8 Most surfaces exhibit mixed reflection containing both diffuse
Lambertian reflection and specular reflection.

no range data) unless the surface normal lies in the epipolar plane and
bisects the projector-target-camera angle. This problem can be reduced by
using cameras with controlled field of view (Nitzan et al. 1987). In a
worse situation, the incident beam may undergo multiple reflections thus
giving false range values.

Structured light systems have certain advantages over stereo vision
systems when seam tracking during welding. These include, minimized
effect of arc illumination and the ability to provide a controlled and steady
intensity illumination of the surface in a range acceptable to the sensor.
Therefore, it is not surprising that this sensing method has by far received
the most attention and is also presented in much detail in this monograph.

2.3.1.4 Time of Flight: This sensing technique uses a signal transmitter
and a signal receiver to collect the signals reflected from the surface being
measured and electronics to compute the roundtrip flight time. The
surfaces being measured should have lambertian reflection characteristics.
These systems can be developed using two types of transmitted signals—
ultrasonic and laser. Time-of-flight systems can be based on one of the
following principles: amplitude modulation phase shift, pulse time delay,
and frequency modulation beat.

18

Amplitude modulation phase shift system modulates the amplitude of the laser beam and compares the incident beam and reflected beam amplitudes to compute the phase shift. The phase shift corresponds to the travel time and hence the distance to the target. This system gives ambiguous results if the phase shift between the incident and the reflected beam is greater than 2π since the range is given by

$$r = n\lambda + r(\phi) \qquad n = 0,1,2,... \qquad (2.2)$$

where

    λ = wavelength of the modulation frequency

    $r(\phi)$ = range as a function of the phase shift for $0 \leq \phi \leq 2\pi$

Pulse Time delay systems directly measure the time of flight using a pulsed laser beam. These systems, however, require the use of sophisticated electronics. Frequency Modulated Beat systems use "chirps" of laser waves that are frequency modulated as a linear function of time, and measure the beat frequency which is proportional to the time of flight (Goodwin 1985).

Unlike triangulation-based systems, the time-of-flight systems do not suffer from shadow regions since the transmitter and receiver can be mounted coaxially thus reducing the baseline substantially without losing measurement accuracy. Ultrasonic signals, however, suffer from lack of resolution required for sensing joint geometries due to the difficulty in generating a narrow acoustic beam. In particular, ultrasonic signals with wavelength smaller than 1 cm attenuate rapidly in air thereby reducing their effective range of use. Additionally, time-of-flight systems based on either ultrasonic signals or laser, suffer from the problem of multiple reflections (similar to triangulation systems), giving greater than correct range.

2.3.2 Indirect Range Measurement Techniques

These techniques infer the range of a target from analysis of the monocular images. As a result, they suffer from long measurement times making them unsuitable for realtime seam tracking applications. The discussion here about indirect measurement techniques is therefore only for informational purposes. Range data can be obtained indirectly through four methods: range computed through focussing, range from images of surfaces with known geometry, range from shadows, and range from

19

Moiré fringe patterns (Nitzan et al. 1987). The method of focussing relies on thin lens optics. If a particular image is in sharp focus then the range can be computed using the formula,

$$r = \frac{s \cdot f}{(s - f)} \qquad (2.3)$$

where,

    r is the distance of the object from the lens center,

    s is the distance of the image from the lens center,

    f is the focal length of the lens.

This method works best if the image of the surface is nonhomogenous with varying intensity features.

For an object with known geometry, the scale changes in the image can be related to the range and thus the range of the various points on the object can be computed. Usually, if the position and the orientation of the object is known then the range of the various other points can be computed using coordinate transformations.

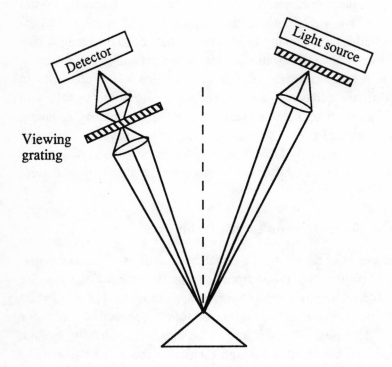

Figure 2.9 Principle of Projection Moiré.

The method for range computation based on measuring dimensions of shadows (Tsuji 1983) is limited to thin objects since solid 3D objects would self-occlude most object points from being seen by both the source and camera. This method has a potential for measuring range discontinuities in stacked objects which are flat (Nitzan 1987).

Moiré fringe patterns give the relative distance between the various equidistant fringes and hence require some knowledge of the object geometry to get position and orientation of the surface being viewed. The fringes can be created by illuminating the surface through a grating and viewing the illuminated surface through an identical grating as illustrated in Figure 2.9.

2.4 EVOLUTION OF SEAM TRACKING STRATEGIES FOR VISION-BASED SYSTEMS

In Section 2.3, we presented an overview of the various possibilities based on through-the-arc and vision sensing techniques. This section now presents an overview of the various seam tracking strategies for vision-based systems. These include the two-pass approach of the first generation seam tracking systems, the one-pass approach characteristic of the realtime second generation systems, and finally, the operational requirements of an intelligent realtime third generation seam tracking system.

2.4.1 Two-Pass Approach (First Generation Systems)

This technique surveys the seam along a pretaught path before welding. Several two-pass systems exist and are based on different concepts. One system for welding seams made of piecewise linear segments (Kreamer, Kohn, and Finley 1986) uses the following approach:

1. Scan each segment at two locations,
2. Compute the coordinates of the root at these points, and
3. Develop a linear model of the segment from the pair of consecutive points.

This being done for all segments in the seam, the points of intersection of adjacent segments are computed from the linear model. These intersection points describe the path along which the torch is guided in phase two. Enhancements to this system include the use of structured light

21

involving angled-camera/light-source for producing two-dimensional video images of the three-dimensional joint shape.

Another two-pass system takes actual images of the joint at the beginning of the sequence (Pavone 1983). The actual image is compared to a template image for the cross-correlation value. The template image is shifted relative to the actual image through the entire range and then the highest cross-correlation value is selected. The shift is measured in the relative x and y directions and the corrections are sent from the image-processor to the robot controller for use in pass two. The system makes correction in two directions, namely, across the joint and in distance from the joint.

Yet another system, quite similar to the one discussed above, uses algorithms for computing the key features of specific joint geometries (Moss 1986). For example, in the case of a vee-joint, these features include the root and the edges of the joint. During the *vision-teach* phase, points are taught along the seam where scanning is to be performed. Subsequently, during the first pass of the tracking phase, the system scans the seam at these points, and processes this data to self-teach the seam characteristics. The seam is welded during pass two by moving the torch along the pathpoints generated in phase one.

There are two advantages to presurveying the workpiece before welding: The first is better control of object illumination and second, the interpretative device/algorithms are not required to work in realtime. Complicated interpretation of sensor signals often takes a long time in comparison with the process feedback control requirements of seam tracking. However, there are some distinct disadvantages in the two-pass method. First, the time taken for presurveying must increase the production cycle time. The other more significant disadvantage is associated with the movement of the joints during welding due to the thermal stresses generated from the intense heat of the welding arc. For these reasons, there is considerable interest in optical sensing approaches which can function during the weld cycle.

2.4.2 Realtime Seam Tracking (2nd Generation Systems)

The operational requirements of such a system include interpreting the sensed data in realtime to compute the joint geometry. This process is further complicated by the presence of arc light and spatter during welding, which adds noise to the sensed data. Most existing systems are designed around these two aspects of realtime seam tracking, as discussed in the following sections. The various seam tracking strategies used are also presented.

2.4.2.1 Enhancements in Hardware. To counter the arc noise, Niepold (Niepold 1979) avoided the arc-light fluctuations in short-circuiting Gas-Metal Arc (GMA) CO_2 welding by synchronizing the camera shutter with the natural short-circuit frequency of the welding process. The optical shutter was opened only during the short-circuit phase and was electronically controlled to avoid both, double exposures and exposures outside the short-circuit range. This prevented the arc light and spatter from adversely affecting the sensed data.

In an alternative approach, Linden and coworkers (Linden and Lindskog 1980) ensured that measurements were not disturbed by the Gas-Tungsten Arc (GTA) welding arc through controlled pulsing of the wirefeed rate, so that the background wirefeed rate was zero and data could be obtained between pulses.

Several researchers have attempted to solve the problems induced by arc light by positioning a mechanical barrier between the arc and the region being observed (Verdon, Langley, and Moscardi 1980; Westby 1977). Heat-resistant brushes made of metal strands, or glass fibers enable the contour of the plate to be viewed and interpreted. Two problems are presented by this technique. The first is the increased distance between the area being viewed and the welding head, which could lead to tracking errors. The second problem arises due to wear of the barrier in production service.

The optical noise introduced by the arc in unshielded systems can also be reduced relative to the sensor signal by use of a strong illuminating source and/or by surveying that area where the arc is weak (Drews and King 1975). At sufficient distance in front of the weld pool there is limited optical interference but there are undesirable time delays between sensing and welding. Simple unshielded systems can be expected to function satisfactorily provided the weld seam is within the detector's field of view. Without any recognition capability, these systems can be confused if the weld seam ever gets outside of the field of view.

2.4.2.2 Pattern Recognition Algorithms. Other approach used by researchers to counter the arc noise is to use some kind of pattern recognition algorithm to convert the image into black and white zones and then extract the widths and position of the desired features. Approaches based on structured light require more complex algorithms to produce shape information from distortion of the structure (Arata, et al. 1975). Increased sophistication in the recognition systems however, allows more tolerance in joint preparation, torch positioning, and process parameters. The system controller should not only be able to interpret the various input signals but also be capable of comparing its present action with the overall welding plan.

2.4.2.3 Seam Tracking Strategies. Initial second generation systems based on visual sensing were limited to tracking straight/circular segments

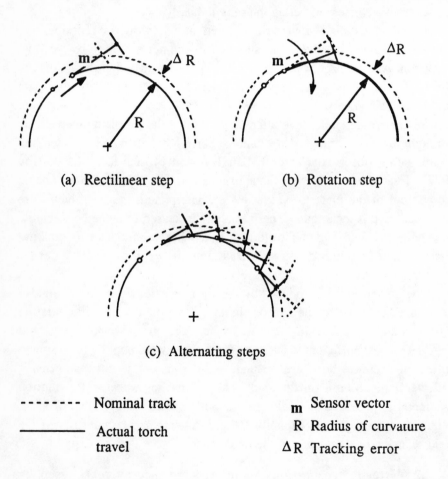

(a) Rectilinear step   (b) Rotation step

(c) Alternating steps

- - - - - - - - Nominal track   **m** Sensor vector

———————— Actual torch travel   R Radius of curvature

ΔR Tracking error

Figure 2.10 Tractrix algorithm movement of the sensor vector following a circular track. (From Lane 1987)

24

with possibly sharp corners (Huber 1987). The operation of two such seam tracking strategies is presented below.

Tracking Planar Segments: If it is known *a priori* that the seam has a large radius of curvature then the following repetitive procedure may be applied:

Step 1: Move the torchtip one step along the direction of the sensor vector **m**, which extends from the torchtip to the root in the current range image.

Step 2: Rotate the sensor to make the center of the range image coincide with the root of the seam. Go to step 1.

This procedure is illustrated in Figure 2.10 and is called the *tractrix* algorithm. The tracking error is negligible when the motion of the torchtip in step 1 is much smaller than the norm ‖ **m** ‖ which in turn must be significantly smaller than the seam's radius of curvature **R**. An advantage of the *tractrix* algorithm is that the *Torch-to-Scanner* coordinate transformation is not required nor are the pathpoint coordinates. The only required information is the rotation of the vector **m** in the torchtip's coordinate frame.

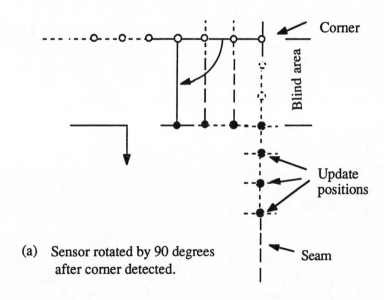

(a)   Sensor rotated by 90 degrees
      after corner detected.

Figure 2.11(a) First approach to tracking seams with sharp corners. (From Lane 1987)

25

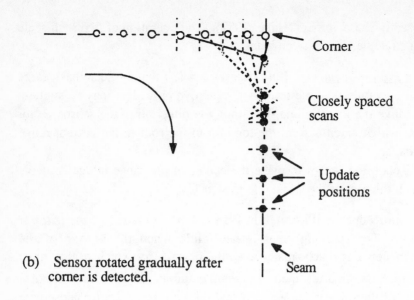

(b) Sensor rotated gradually after corner is detected.

Figure 2.11(b) Second approach to tracking seams with sharp corners. (From Lane 1987)

Another procedure called the *side-step* algorithm, tracks straight segments without rotating the sensor. The sensor, attached to the torch, is placed so that the seam is within the scanning window. The sensor vector **m** is positioned roughly parallel to the general direction of the seam, thus ensuring that the straight seam always remains within the scanning window. The coordinates of the pathpoint are calculated from both, the range image and the sensor vector **m**, and stored in the robot control module.

Tracking sharp corners: If the scanwidth is large compared to the seam cross-section, sharp corners can be detected as an abrupt change in the profile. As soon as this occurs, the sensor can be rotated about the torch to track the seam beyond the corner using one of the two possible approaches illustrated in Figures 2.11(a) and 2.11(b).

First Approach: On locating a sharp corner and knowing the direction in which the seam turns at that corner, the sensor can swing a complete 90 degrees while the torchtip still keeps moving the length of the sensor vector. As the torchtip reaches the corner, the sensor's scanning window moves over the seam and the system can track the seam beyond the corner. It should be noted that a small

26

section of the seam beyond the corner is tracked blindly because the sensor is rotated by 90 degrees immediately after locating the corner.

Second Approach: This requires that when a corner is located, the sensor rotation proceeds in a controlled fashion so as to keep the scanning window over the seam. When the rotation of the sensor reaches approximately 90 degrees, the straight line tracking beyond the corner can now commence. It should be noted that the points scanned immediately after the corner are no longer equidistant and hence, coordinates of the intermediate path-points have to be interpolated.

2.4.3 Third Generation Intelligent Systems

Third generation systems have higher operational requirements than those for existing realtime systems. In addition to interpreting noisy sensor data in realtime, these systems are required to have the following features:

- The system should sense and react to changes in the global environment besides sensing the seam.
- The active vision sensor should be self-adjustable to a broad spectrum of arc intensities.
- The system should adaptively control the welding parameters depending on the seam environment.

Figure 2.12 illustrates schematically the general principle of the third generation systems whose functions include flexible seam tracking adapted to most weld preparations, online quality control of bead shapes, and detection of downstream obstacles.

Control systems for controlling complex activities in unstructured environment such as the seam, can benefit from combining concepts from artificial intelligence (AI), realtime systems, and control theory to provide _Intelligent Realtime Control_ (IRTC). IRTC is appropriate whenever time-critical decision-making functions are required within a control system. In any realtime intelligent system, there is a fundamental tradeoff between acting and reasoning. Time is a valuable resource that is lost if the system must reason about the action before performing them. In some cases, failure to act can be the worst possible action. On the other hand, reasoning is necessary for proper interaction with the environment and can often avert future delays or disasters.

27

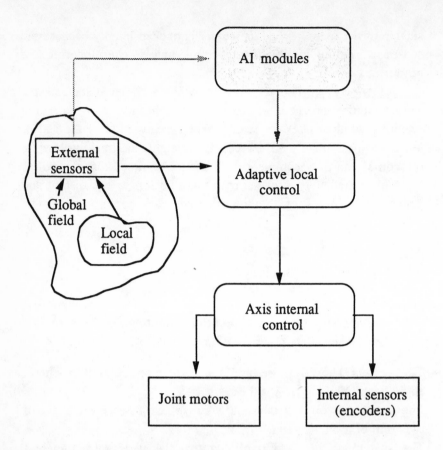

Figure 2.12 Conceptual overview of third generation seam tracking systems.

Adaptive, Realtime, Intelligent Seam Tracking. IRTCs in the context of seam tracking for robotic welding are the subject of this monograph. Readers should note that the welding process control aspect of these systems is outside the scope of this monograph which primarily addresses issues related to seam tracking. To perceive the nature of these third generation seam tracking systems, it is helpful to understand their relevant attributes.

Seam Tracking: The system should track a seam that may change in all three dimensions. The seam tracking algorithms discussed so far are based on correcting the tracking error in 2D. However, a more robust seam tracking algorithm is required for tracking curved seams in 3D space. The

28

approach presented in this monograph computes the path-points (control input) to correctly position and orient the torch and the lookahead sensor over the seam from an adaptive model of the seam geometry (details of this method are provided in Chapter 5). The seam tracking algorithm should also ensure that the tracking error does not propagate from one cycle to the next, thereby establishing an upper bound on the tracking error (details provided in Chapter 7).

Adaptive: This represents the system's ability to track a seam along non-preprogrammed weld paths. As in any typical adaptive process, the precise description of the plant (seam) is unknown. In the absence of an *a priori* known weld path, a model of the seam is assumed and its parameters are adaptively computed. The geometric model of the seam is generated from the world coordinates of the seam's feature points identified in the range image.

Realtime: In a realtime system, the specifications for correct system behavior contain constraints related to time in the real world. These constraints may be explicit functions of time or may be implicit in some other constraints (for example, tracking a certain seam at a specified speed and within allowable tracking error). In realtime seam tracking, the system simultaneously learns about the seam while tracking it. It also interacts with the seam's environment in realtime. The realtime requirement therefore, constrains the maximum tracking speed for a given radius of curvature and vice versa. This constraint is necessary to keep the tracking error within desired tolerances.

Intelligent: Intelligence in control activities represents the capability of the system to respond to its environment autonomously. In traditional realtime systems, the basic requirement is to provide a faithful transformation of the externally provided control signals into physical actions. Intelligent systems, however, are capable of performing complex actions with minimal external guidance in response to the sensed environment.

In an unstructured environment such as the seam, the various possibilities the system may encounter are well-defined, but their actual sequence of occurrence is not. The system should be able to recognize the seam environment through extracting and interpreting the seam's characteristic features from noisy and distorted range images (details provided in Chapter 3). It should make realtime decisions about interacting with the seam environment based on this global information.

29

2.5 SUMMARY

The two most widely used sensing techniques used in seam tracking include, through-the-arc sensing and vision-sensing. Through-the-arc systems sense the arc signal itself to recognize the seam geometry. Although the sensed data is local in nature, through-the-arc systems based on various seam tracking algorithms have been developed. Depending on their hardware configurations, they may be classified as general purpose systems, robotic bolt-on hardware types, or integrated robotic through-the-arc systems.

Vision sensing has been extensively studied and applied to seam tracking problems. It provides the advantages of sensing the environment outside the welding process but suffers from the arc light noise. Range measurements from vision sensors fall under the category of direct ranging and indirect ranging. Direct ranging systems are typically used for seam tracking applications and amongst them, the scanning beam laser range sensors and the structured light projector/camera range sensors have received considerable attention. These systems have been presented in detail in this monograph.

The evolution of vision-aided seam tracking systems has seen three stages: the first generation two-pass systems with separate self-teaching and welding phases; the second generation realtime systems that correct the preprogrammed path and weld the seam in the same pass; and finally, the third generation intelligent systems that track and react to an unstructured seam environment in realtime. The last is the main subject of this monograph.

CHAPTER 3

FEATURE EXTRACTION AND SEAM RECOGNITION TECHNIQUES

3.1 INTRODUCTION

The previous chapter discussed the various techniques available for sensing and measurement of the surface geometry. This chapter provides details on processing the vision data for extracting and recognizing the seam's features in the image and subsequently modeling the seam environment. In a broad sense, the images generated as a result of the signal transduction by vision-based systems can be classified as intensity images or range images. Intensity images are a rectangular array of intensity values corresponding to spatial points on the surface being measured. On the other hand, range images generated using position sensitive photo-detectors can be described by an array of range values (distance) corresponding to spatial points in the scene being measured. These range image arrays are rectangular when the laser beam can be moved along two axes and linear in the case of synchronized scanning systems where the beam is scanned along a line.

For seam tracking applications, the final form of the sensed data should describe the joint geometry and its location in either the world space or in relation to a known position of the torchtip. Thus, it is necessary to transform the sensed data from the image space to world space during some stage of the process. Since transformation to the world coordinates is computationally expensive, it is done only in the final steps of the processing scheme. In the case of structured light projection, the intensity image is processed in pixel coordinates before computing the range using triangulation principles. Range image generated by synchronized scanning of a laser beam along a line has the advantage of describing a profile in a fewer number of rangels, thus allowing transformation of the entire range image if required.

The data obtained in intensity as well as range images suffers from errors affecting both, the value of the measured data and the image coordinates of the spatial points measured. Hence, signal preprocessing is usually the first step in image processing. This step results in an improved image in which the errors are minimized. Signal preprocessing is followed by feature extraction wherein characteristic features describing the object being measured are identified and located in the processed image. In the case of welding seams, these features include seam edge points, sloped surfaces, root points, etc. The interpretation of the extracted features follows next and it allows matching the extracted features with known model features in order to determine the identity and location of the sensed object. In the case of welding seams, feature interpretation provides the system with knowledge about the type of seam and its geometric description for generating a model of the seam. This model of the seam forms the basis for all interaction between the seam tracking system and the unstructured seam environment.

The following sections discuss in detail the various techniques, methods, and issues related to signal preprocessing, feature extraction, feature interpretation, and seam modeling in the context of seam tracking for robotic welding. To provide the reader with a better understanding of image processing in an unstructured seam environment, all discussions are made in the context of a representative vee-jointed seam with tack welds. Possible scenarios describing interaction between the robotic welding system and the seam environment for welding process control are also presented for the sake of completeness.

3.2 INTENSITY IMAGE PREPROCESSING

Intensity images of laser stripes from structured light projectors illuminating the welding seam can be expected to have noise from a variety of sources. For example, welding arc light, spatter, and multiple reflections from specular surfaces can cause the measured data to be erroneous. Additionally, the variety of operating conditions such as varying surface preparations and surface orientations introduce requirements of high operational reliability on the vision sensing for seam tracking systems.

Much work has been done to improve the quality of intensity images for a variety of applications. Using techniques such as Gaussian, median and other weighted filters in both one and two dimensions, noise in the intensity images can be minimized. Other modifications to the intensity images have been designed for special applications. For example, logarithmic transformation of the intensity image enhances the dynamic range in the darker end of the intensity image while histogram equalization spreads the most significant portion of the intensity image (Rosenfeld and Kak, 1982). It should be noted that most filtering operations lose some information due to integration over a wide spatial region. However, the resulting image can be expected to have better signal-to-noise characteristics. Improvement of intensity images for the specific purpose of detecting laser stripes is performed on the basis of either:

1. brightness, using intensity thresholding, or
2. spatial continuity and thickness through spatial filtering.

Intensity thresholding can be performed on contiguous sequences of pixels in the image, either along vertical or horizontal scan lines, by selecting only those pixels whose intensity exceeds a certain threshold. The center pixel in each sequence can then be used in representing the laser stripe image. The effect of weld noise and spatter can be minimized by rejecting all pixel sequences which are greater than a certain size selected as representative of the laser stripe thickness. In another approach, if the laser stripe image is not expected to change drastically from image to image, then a suitable number of consecutive binary images can be overlapped and the effects of flying spatter can be removed with AND operation (Kawahara 1983). However, this approach severely restricts the seam tracking systems ability to track general 3D seams where the laser stripe image coordinates can change from image to image. To compensate for varying background intensity in different portions of the image and also from image to image, the intensity threshold should be adaptively selected. This can be effectively done by analyzing the cumulative intensity histograms on a line and by exploiting the known thickness of the stripe (Agapakis, et al. 1990a). Other approaches include automatically setting the threshold by averaging the analog video signal itself. This design permits the generation of a normal binary image even if the general illumination level of the target surface changes drastically, eg. from 0 to 13000 lux as reported by Kawahara (Kawahara 1983).

Spatial filtering, on the other hand, involves convolution of the laser stripe image with an appropriate filter. The result is an identifiable response along the centerline of the constant thickness laser stripe image, while not reacting to other sources of brightness within the image. Since the intensity profile across a laser stripe can be approximated as a rectangular pulse of a constant width or a Gaussian, a filter such as the Laplacian of a Gaussian in 2D or the second derivative of a Gaussian in 1D can be used. Such a filter can be approximated using the second difference operator (Agapakis, et al. 1990a). The implementation involves the application of a mask of constant

33

width n to the laser stripe image. Typically, the laser stripe image can be expected to be either vertical or horizontal and so the filter (mask) can be one-dimensional. Thus, the mask is constructed with a central region of size n and gain 2 surrounded on either side by regions of size n and gain -1. Furthermore, the continuity of the laser stripe image can be exploited to enhance the performance such that only the area around the previously detected laser stripe is processed instead of the entire image. In case of discontinuities and ambiguities, the processing of the entire window resumes.

The error in image coordinates as encountered in intensity images include, barrelling effect or pincushion distortion, skew, and rotation. These errors can be corrected by comparing a target object with known geometry with its image. Gennery is reported to have used a target with uniformly spaced lines and dots for this purpose (Gennery 1982). This comparison process, known as calibration, can be used to compute the error, which is usually represented in parametric form. Subsequent images of unknown objects can now be subjected to the inverse transformation, known as decalibration, to find the correct coordinates of the image points. However, the process of calibration/decalibration can result in many image points from the original image to be mapped onto the same coordinates in the transformed image, causing the corrected image to be blurred with loss of resolution.

3.3 RANGE IMAGE PREPROCESSING

Range image preprocessing deals with identifying erroneous range values and discarding them. These errors are usually associated with either dark surfaces or specular surfaces. Dark surfaces reflect few photons and hence require a longer integration time. Many ranging systems based on synchronized scanning and using position sensitive charge-coupled devices (CCD) provide a control on the exposure time of the CCD array receiving the reflected beam. To prevent long exposure times from adversely affecting the realtime requirements of the seam tracking system, the cumulative exposure time should be monitored. If the cumulative exposure time is greater than a certain limit while the signal value is less than a certain threshold then it is assumed that data is missing for that spatial point.

Just as low exposure for dark surfaces can result in missing data, high exposure on bright surfaces can cause undesirable "blooming" of the CCD array, resulting in incorrect range values. The range image acquisition time

34

per exposure cycle is a function of the exposure counts and the fixed CCD transfer time. Since an unsatisfactory signal peak means repeating the exposure cycle and consequently increasing the range data acquisition time, the exposure count should be adaptively selected for every exposure cycle. This can be effectively done by examining the exposure count and the signal strength at the corresponding scan position during the previous scan and also the average exposure count during the previous scan. In addition to the CCD exposure control, the CCD array should be calibrated to mitigate the effects of ambient light and provide a good resolution for the CCD output signals. Appendix C at the end of the Chapter 6 discusses the setting of the voltage levels during the calibration process of a scanning beam laser profiling gage.

Specular surfaces present considerable difficulty to the operation of the structured light ranging systems. Unless the normal to the specular surface bisects the angle formed by the incident beam, the target point, and the reflected beam, the CCD camera would not be able to receive the reflected

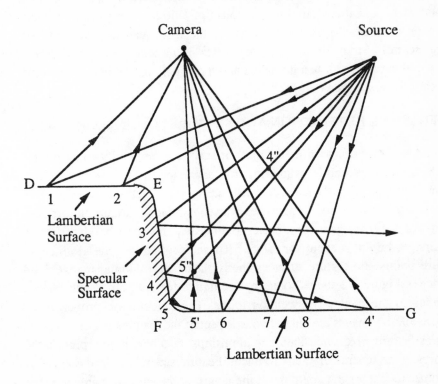

Figure 3.1 Effect of multiple reflections on triangulation-based ranging.

35

beam, thus, resulting in missing data for the spatial point. Besides this problem of the reflected beam not reaching the camera, specular surfaces can produce incorrect range values due to multiple reflections of the beam before reaching the camera as shown in Figure 3.1. This problem is quite common in the case of brushed steel or aluminum surfaces with vee, lap, and fillet joints.

The error due to multiple reflections is usually detected using heuristics based on model of the seam geometry. If the range data for a particular scanning position of the beam does not match the expected trend for the entire scan, then such range data can be regarded as erroneous and should be either measured correctly or ignored. While tracking vee-jointed welding seams, the geometry of the seam causes any multiple reflections on the CCD array to reduce the true range value. The heuristic for bad data suppression is based on this knowledge. Another concern while tracking unstructured 3D seams is the frequently changing position and orientation of the sensor in relation to the seam. Since the heuristic for detection of erroneous range values analyses the trend of the range data, it should be robust to handle this changing relationship. One approach removes the effect of the variable relationship between the sensor viewing angle and the target seam by transforming the range data into another coordinate frame located along the seam being tracked and then applying the heuristic on the transformed range data (Nayak 1989).

3.4 FEATURE EXTRACTION AND SEAM RECOGNITION

A feature in the image of a 3D object or surface is an entity consisting of one or more measurable properties that characterize the object or scene being analyzed. Usually features constitute such geometric entities as edge points, surface lines and arcs, angles between surface lines, etc. In order to recognize the seam geometry, corresponding features must be identified in the image, be it intensity or range. These features in general include feature points and feature surfaces (both planar and nonplanar machined surfaces). Figure 3.2 shows some representative feature points and feature surfaces in the context of welding joints. These geometric properties as captured in the images are but only one aspect of the features. Other aspects such as feature size and feature relationships also help in interpreting the features in an unstructured environment. Feature size relates to the value of a geometric feature and could mean the length of the surface segment or the value of the angle between two planar feature surfaces, etc.

36

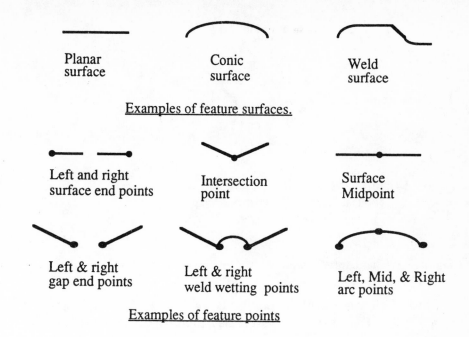

Planar
surface

Conic
surface

Weld
surface

Examples of feature surfaces.

Left and right
surface end points

Intersection
point

Surface
Midpoint

Left & right
gap end points

Left & right
weld wetting points

Left, Mid, & Right
arc points

Examples of feature points

Figure 3.2 Feature surfaces and feature points in the context of welding.

The feature size along with feature relationships can be used in heuristics to interpret the entire image and determine the location of the object or scene being analyzed. In addition to the geometric features provided by range and intensity images, the intensity images also provide photometric features such as intensity histograms and color. These features, however, are more relevant to welding process control rather than seam tracking and hence are mentioned only for the sake of completeness.

In most manufacturing applications, the joint geometry for the various seams can be described in terms of feature attributes and feature relationships. For example, in robotic welding, the common weld joint types encountered are fillet, vee-groove, split, lap, etc. Figure 3.3 shows the definition of the vee-joint in terms of its characteristic features.

In a structured robotic welding environment, all variables can be well-defined *a priori*. The various possible seam types that the sensor would encounter in each image frame, the location and orientation of the sensor in relation to the seam, etc., are known. The image processing algorithms developed for such systems are specific to the application in hand. Example of such a system includes the seam tracking system developed for robotic

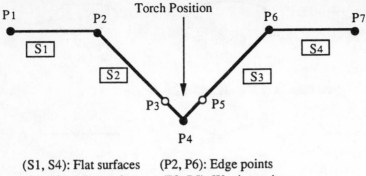

(S1, S4): Flat surfaces   (P2, P6): Edge points
(S2, S3): Joint surfaces   (P3, P5): Wetting points
    P5: Root point   (P1, P7): Scan endpoints

Figure 3.3 Characteristic features of a vee-grooved weld joint.

welding of circular penstocks in tunnels, which has shown good practical results. (Kawahara, 1983). However, the extremely controlled environment in which these seam tracking systems operate do not allow their image processing algorithms to be easily adapted for unstructured welding situations.

Robotic welding systems required for operation in unstructured seam environments have more stringent requirements. Although the various scenarios that the system would encounter during seam tracking are well-defined (eg. fillet joint, slip joint, vee groove, tack welds, end-of-seam conditions, etc.), the sequence of their occurrence along an unstructured seam is not known *a priori*. Furthermore, seam recognition from a noisy range image is a complex process, especially when the image of the seam geometry is expected to have a certain degree of variability (refer to Figure 3.4). Some misalignment between the center of the scanning window and the root of the seam is inevitable during the seam tracking process, leading to the loss of some features of the range image and thereby adding to the many variations in the seam types. Creating a static knowledge base of the various seam types plus their possible variations and using correlation for matching with the current range image is not feasible for realtime operation of the system (Levine 1985; Gonzalez and Winz 1987).

In the context of realtime intelligent seam tracking in an unstructured environment, two approaches for feature extraction and feature recognition are presented: *top-down* seam recognition and *bottom-up*

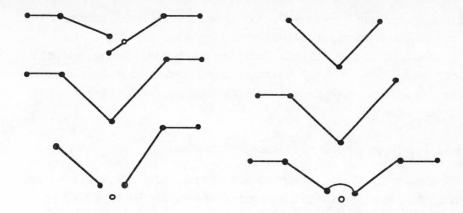

Figure 3.4 Variability in weld seam image instances (including convex corners, tack welds, and partial images).

seam recognition. The top-down approach starts with heuristics to hypothesize the seam type that is most likely to be encountered in the current scan. This is followed by the application of algorithms specific to that seam type to verify the hypothesis and extract the features in the image. Figure 3.5 shows an overview of the top-down approach to seam recognition and subsequent seam environment modeling and interaction. On the other hand, the bottom-up approach models complex seam types from primitive features, such as straight lines and arcs. This process starts with the use of general segmentation algorithms to extract the primitive features, followed by construction of a contour from these primitive features and then matching the contour to the expected contour type. The matching process provides labels for the various extracted features and their location in the image.

The top-down approach works well for realtime seam tracking since the recognition algorithms are designed for a specific seam type. However, as new seam types are introduced, new recognition algorithms have to be designed, developed and implemented in the seam tracker. The bottom-up approach, on the other hand, is data driven. Therefore, no new feature recognition software has to be developed for the recognition of a new contour type although a model of each new seam type has to be constructed in terms of the primitive features and stored in the model library. However, the open architecture of the bottom-up approach can be computationally expensive since good segmentation algorithms used in primitive feature

39

extraction are mostly iterative in nature and detailed representation of the seam contours may require extensive segmentation of the image.

The following sections provide further details on the two approaches to seam recognition. The top-down approach is presented in the context of range image processing while the bottom-up approach is presented for intensity images.

3.4.1 Top-Down Approach to Seam Recognition

The approach used here is based on heuristic reasoning, which allows seam recognition in realtime with a high degree of certainty. By applying a set of heuristic rules, the decision is narrowed down to a particular seam type, which is followed by the extraction of relevant features in the image using seam-specific segmentation algorithms. The resulting features are subsequently examined using verification models specific to that seam type. This strategy, in most cases, obviates the need for detailed pattern matching between the given range image and all the possible seam type models stored in the static knowledge base.

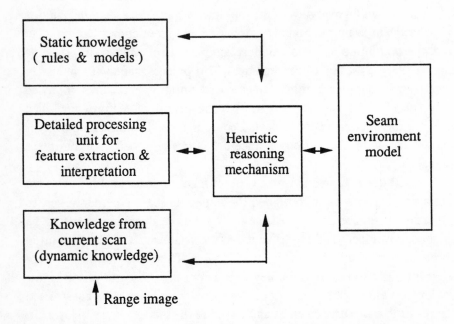

Figure 3.5 Overview of the top down approach to seam recognition and seam environment modeling.

The following sections discuss in detail, the recognition algorithms applicable to slip-joints, fillet joint, tack welds, and end-of-seam condition. Sharp corners along the seam are also treated as end-of-seam condition by the system. Tack welds and the end-of-seam condition are recognized by analyzing a sequence of images rather than just a single image. In this case, the system does not immediately conclude about the seam type, but generates a *presently unknown* seam type, which is later classified into either a tack weld or the end-of-seam condition.

Seam recognition is performed in four steps. In the first step, the range image is evaluated using simple heuristic rules resulting in an *object-attribute-value(OAV)* triple for each possible seam type that the system may encounter along the seam. This *OAV* triple, called *cur_scan_cntxt*, contains the dynamic knowledge regarding the seam type, and can be explained in *OAV* terms as:

object: possible *seam type* associated with the current scan.
attribute: *certainty factor* of this particular seam type being true in the current scan context.
value: a real number between +1 and -1 indicating the system's degree of confidence in the above attribute being true or untrue.

In the second step, a set of heuristics are applied based on the current values of the dynamic knowledge in *cur_scan_cntxt*. The intention is to narrow the scope of the various possibilities to establish the most likely seam type. The value of the *certainty factor* is adjusted for all possible seam types depending on the outcomes of the rules applied. If this process is represented by a search tree, then the continuous adjustment of the *certainty factors* ensures an intelligent search for the most likely seam type.

The third step involves the application of seam-specific algorithms to locate the seam's characteristic features in the range image. Verification of the seam type's validity, performed in step four, is based on matching the extracted features with the corresponding seam type model stored in the system's static knowledge base.

The above approach using heuristic reasoning has two advantages: First, a detailed pattern matching process is not necessary for every range image scanned, since simple heuristic rules help in narrowing down the various possibilities; Second, the verification process, which is actually model-based pattern matching, can be accomplished using relatively fewer details in the static knowledge base. This is possible because the previous

41

decision based on heuristic reasoning establishes a high degree of confidence in the seam type associated with the current range image, before passing it to the verification module.

3.4.1.1 Implementation of the Top-Down Approach. The implementation of this seam recognition scheme is a four step process as described below. This particular implementation is designed for processing range images of vee-joints and has the following steps.

1. Initial range image evaluation,
2. Most likely seam type classification,
3. Feature extraction using seam-specific algorithms, and
4. Verification of the extracted seam type.

<u>Step 1:</u> Initial range image evaluation starts with a filtered range image from which spurious range values, if any, have been removed using a heuristics mentioned in Section 3.3. The output of the range image evaluation module is an *OAV-triple* representing the certainty regarding the current range image's seam type. This certainty is quantified to a value between +1.0 and −1.0, reflecting the system's degree of confidence in its belief. A value of +1.0 indicates that the system believes the hypothesis is true, while −1.0 is associated with the hypothesis being untrue. In the development of the rules, the values assigned to the certainty factors are empirical. The steps involved in evaluating the range image are:

1. Find the average range in the range image (*cur_avg*).
2. Apply the following rules:

Rule 01: If the absolute difference between the average range (*cur_avg*) and range of the first point in the range image is less than a threshold value, then

> *certainty factor(vee joint) = -1.0*
> *certainty factor(tack weld) = 0.0*
> *certainty factor(undefined seam) = 0.0*
> *certainty factor(presently unknown seam) = +1.0*

Remark: The value of the threshold is selected by the system based on the expected size of the seam. ♣

Rule 02: If Rule 01 is applicable, and if the *cur_avg* is less than the first range data, then

$$certainty\ factor(\ tack\ weld\ ) = 0.5$$

Rule 03: If Rule 01 is not applicable, then

$$certainty\ factor\ (\ vee\ joint\ ) = +0.8$$
$$certainty\ factor\ (\ tack\ weld\ ) = -1.0$$
$$certainty\ factor\ (\ undefined\ seam\ ) = -1.0$$
$$certainty\ factor\ (\ presently\ unknown\ seam\ ) = -1.0$$

<u>Step 2:</u> The module for establishing the most likely seam type further narrows the scope of the possible seam types based on the *certainty factor* values in the dynamic knowledge base (*cur_scan_cntxt*). The rules used in this process are

Rule 11: The seam type of the current range image is most likely to be the same as that of the previous range image. If the *certainty factor* corresponding to the previous range image's seam type is greater than 0.5, then the possible seam type for the current range image is same as the previous range image.

If *Rule 11* is not applicable, then the possibilities are:

Rule 12: The scanning has crossed from a region whose seam type is *presently unknown* to a normal weld seam region. Hence, the previous undecided segment can now be identified as a *tack weld*. Also, the seam type in the current range image is most likely to be the same as that in the region before the *tack weld*. If the *certainty factor* corresponding to the seam type in the previous segment is greater than 0.5, then the possible seam type of the current range image is the same as that of the previous segment.

Rule 13: The scanning has crossed from a normal weld seam into a tack weld region. If the *certainty factor* corresponding to *presently unknown* seam type is greater than 0.5, then decision regarding the current range image's seam type is not made immediately since it could be either a *tack weld* or the *end-of-seam* condition.

Rule 14: If Rule 13 is applicable and the seam types associated with all scans between the current torch location and the current scan are of *presently unknown* seam type, then it is the *end-of-seam* condition.

Remark: For reliability in identifying tack welds through the feature interpretation algorithm, the length of the tack should be less than the distance between the torchtip and the beam. This assumption is practical

43

for preassembling the weld parts and is treated as a specification for seam preparation prior to welding.   ♣

<u>Step 3:</u> Based on the lead of a possible seam type, seam features are extracted using algorithms specific to the seam type in the current scan. Discussed here are algorithms specific to vee, split, and fillet joints. Similar algorithms can be developed for extracting features from other seam types.

Feature Extraction Algorithm for Vee-Grooves: In the digital image processing terminology, the process of partitioning the image space into meaningful regions is known as segmentation. In this case, the meaningful regions refer to the seam's features. The features of a vee, split and fillet joints are as shown in Figure 3.3 and the corresponding features as seen in a range image are shown in Figure 3.6. The segmented range image of a vee, split, and fillet joints are characterized by three feature points: the two outer points describe the edges while the center point describes the root of the joint. For the fillet joint, the two outer points correspond to the first and the last data points in the scan. Other features such as surfaces and angles, and the feature sizes can be computed from the feature points.

The feature extraction algorithm for vee-joint presented here is a modification of a vee-groove detection algorithm by Smati, Smith and Yapp (Smati, et al. 1984) and assumes a *segmentable* range image. A *segmentable* range image of a vee-joint has all points on the top flat surfaces,

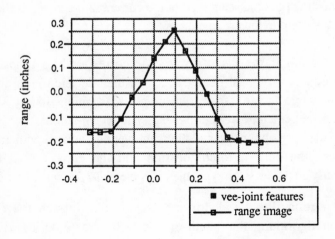

Figure 3.6 Characteristic features of a vee-joint as seen in the range image.

44

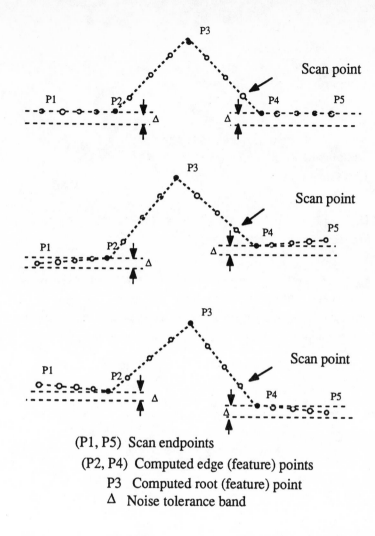

(P1, P5) Scan endpoints
(P2, P4) Computed edge (feature) points
P3 Computed root (feature) point
Δ Noise tolerance band

Figure 3.7(a) Acceptable variations in vee-groove range images for
feature extraction (segmentable range images).

i.e. region outside the vee, within a specified tolerance band associated with
the joint.

Figure 3.7(a,b) illustrate examples of *segmentable* and *nonsegmentable*
range images. Since the position of the sensor in relation to the seam is
expected to change during seam tracking, the original range image may not
necessarily be segmentable. Hence, it has to be transformed into a suitable

45

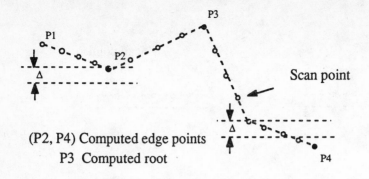

(P2, P4) Computed edge points
P3 Computed root

Figure 3.7(b) Unacceptable vee-groove range image for feature extraction
(nonsegmentable range image)..

coordinate frame to make it segmentable. Details of this image
transformation process are discussed in Chapter 5.

The feature extraction algorithm is based on an iterative averaging
technique and the pseudocode of the algorithm for Vee-joint is as follows.

First Average.

$$\overline{Y}_0 = \frac{1}{N}\sum_{i=1}^{N} Y_i$$

Find   L   s.t. $Y_L > \overline{Y}_0$, $Y_{L-1} < \overline{Y}_0$, $1 \le L \le N$

Find   R   s.t. $Y_R < \overline{Y}_0$, $Y_{R-1} > \overline{Y}_0$, $1 \le R \le N$

Left Edge:

$$\overline{Y} = \overline{Y}_0$$

Repeat

   1. If ($Y_k < (\overline{Y}$ - tolerance)),   k = L ... 1

     Then   L = k

   2. $\overline{Y} = \frac{1}{L}\sum_{i=1}^{L} \overline{Y}_i$

Until $Y_k \ge (\overline{Y} - tolerance)$, $\forall k \in [1, L]$

Left edge = L

46

Right Edge:

$$\overline{Y} = \overline{Y}_0$$

Repeat

    1. If $(Y_k < (\overline{Y} - tolerance))$, $k = R \ldots N$

    Then $R = k$

    2. $\overline{Y} = \dfrac{1}{(N - R + 1)} \sum_{i=R}^{N} Y_i$

Until $Y_k \geq (\overline{Y} - tolerance)$, $\forall k \in [R, N]$

    Right edge = R

Root:

    Find k *s.t.* $Y_{k+1} < Y_k$, $k = L \ldots R$

    Root = k

Remark: To remove the effect of minor surface indentations in the range image, a lowpass filter is used (details provided in Section 5.3.3 of Chapter 5). The tolerance band is based on the size of the seam and the resolution of the scanning beam positions in the scan.   ♣

<u>Step 4:</u> The model-based verification in step four verifies the decision made through heuristic reasoning regarding the possible seam type of the current range image. The verification process is essentially geometric reasoning (Clocksin, et al. 1985), i.e., matching a model computed from the current range image with predefined models of possible seam types. These models, for purposes of verification, are represented in terms of salient features of the seam geometry. For a vee-joint, these include the length of the left and right top-edges, the left and right bevel-edges, and the included angle between the bevel-edges. Similar models can be derived for other seam types. If the output of the verification module does not match the most likely seam type identified in step two, then an error condition is generated requiring user intervention.

Rule 21: Model-based verification of vee and slip-joint (refer to Figure 3.3)

1. Total number of edges should be at least 3,
2. Sum of the left and right top-edge lengths should exceed a specified threshold ($s_1 + s_4 >$ threshold),
3. Length of the left bevel-edge should exceed a specified threshold ($s_2 >$ threshold),
4. Length of the right bevel-edge should exceed a specified threshold ($s_3 >$ threshold), and
5. Included angle should be between certain limits.

Rule 22: Model-based verification of fillet joint (refer to Figure 3.3)

1. Total number of edges should be 2,
2. Sum of the left and right bevel-edge lengths should exceed a specified threshold ($s_2 + s_3 >$ threshold),
3. Length of the left bevel-edge should exceed a specified threshold ($s_2 >$ threshold),
4. Length of the right bevel-edge should exceed a specified threshold ($s_3 >$ threshold), and
5. Included angle should be between certain limits.

Remark: The threshold values used above are empirical and specified in the description of the geometric models for the various seam types ♣

3.4.2 Bottom-Up Approach to Seam Recognition

This approach allows the weld joint to be specified as a collection of primitive features which are then recognized in the image. This approach typically consists of three steps (Agapakis, et al. 1990a):

1. contour preprocessing;
2. contour shape representation;
3. contour feature recognition.

The detailed structure of the bottom-up approach to seam feature recognition is illustrated in Figure 3.8.

3.4.2.1 Contour preprocessing. This step is similar to the signal preprocessing discussed in Section 3.2. In this step the intensity image is prepared for detailed analysis. Any gaps in the image are identified and a gap filling operation allows continuous representation of the image. Algorithms for fast line drawing as used in computer graphics applications

48

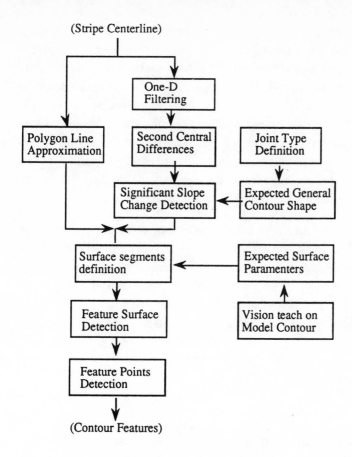

(Stripe Centerline)

One-D Filtering

Polygon Line Approximation

Second Central Differences

Joint Type Definition

Significant Slope Change Detection

Expected General Contour Shape

Surface segments definition

Expected Surface Paramenters

Feature Surface Detection

Vision teach on Model Contour

Feature Points Detection

(Contour Features)

Figure 3.8 Details of the bottom-up approach to seam feature recognition.[†]

can be used to perform this function. Gap filling is followed by filtering the image to remove noise.

3.4.2.2 Contour shape representation. In the unstructured environment of the seam with frequent changes to the location and orientation of the light stripe and the seam in the image, it is necessary to have a robust approach to represent the weld seam geometry. One method uses multiple representations of the same contour such that they complement each other

---

[†] Reprinted from Int. Journal of Robotics Research, Vol. 9:5, 1990, Agapakis et al., Approaches for recognition and interpretation of workpiece surface features using structured lighting, by permission of the MIT Press, Cambridge, Massachusetts, Copyright 1990 Massachusetts Institute of Technology.

(Agapakis, et al. 1990a). The joint geometry is represented both in terms of its feature points and feature segments (feature surfaces). Feature points are computed based on detecting significant orientation changes along the stripe image. In a general approach, the segments that are identified can be either straight or curved. However, most weld seam feature surfaces if machined or flame-cut can be expected to be straight although curved surfaces can be equally well modeled. Figure 3.2 shows some typical feature surfaces and feature points that could be encountered during robotic welding. The following sections discuss the methods for locating the feature points in the images and for modeling the feature surfaces.

Feature points. These can be located by detecting orientation changes at those points along the image where the slope changes abruptly between two line segments. Techniques such as second difference operation (Agapakis, et al. 1990a) or detection of the positive-to-negative zero crossing in an image filtered using a least mean square slope detection filter (Corby 1984) can be used to locate feature points along an image. The second central difference operator is applied along a moving window of size $2\,k$ and is of the form

$$s(u) = \frac{v(u+k) + v(u-k) - 2v(u)}{2k} \tag{3.1}$$

where $v(u)$ are the coordinates of the laser stripe image when represented as a single valued profile.

On a straight line segment, the second difference operator response is zero. A nonzero response at any point (u) along the profile describes a situation where the slope has abruptly changed. The window size k controls the amount of smoothing introduced before the differences are calculated. A smaller value of k results in high sensitivity to local or minor changes which can occur due to noise or minor surface irregularities while larger values of k allow detection of orientation changes which are more global in nature. Hence, the window size should be selected on the basis of the expected stripe geometry which in turn is dependent on the geometry of the weld seam being tracked.

The feature points detected as orientation changes provide a coarse representation of the stripe. This model of the weld seam has the advantage of simplicity and can be used to compute the torch position and orientation during seam tracking. However, this model is more susceptible to errors due to resolution of the camera and quantization errors. A more detailed representation of the stripe can be provided by modeling all the feature surfaces as well.

Feature surfaces. This model provides the description of the weld seam in the areas bounded by the feature points. The importance of having a dual model representation of the contour shape in terms of both feature points and feature surfaces is in improving the accuracy of the previously calculated feature point locations.

Modeling the feature surfaces involves segmenting the entire stripe image in terms of straight line segments and arcs. Most heuristic based schemes are quick and provide good segmentation results. Several methods have been proposed to segment a sequence of points into straight lines. A few methods propose to segment curves into arcs (Albano 1974; Liao 1981; Pavlidis 1982b). Unfortunately, higher order curve fitting is difficult since the curves may not model the region between the set points. So an approach using a set of simpler curves such as circular arcs to model the original curve is preferred.

Most weld joint surfaces prepared for robotic welding are machined or flame-cut and so their feature surfaces in the stripe image can be represented by straight lines. Hence, segmentation algorithms designed for representing contours using straight segments provide better performance. However, in the interest of generality, we also describe here a segmentation algorithm which represents general contour shapes using both straight line and curved segments. In the context of multipass welding, this algorithm can be used to model weld bead geometries although this level of detailed processing has not yet been reported in the literature. Two algorithms for segmenting general contour shapes are presented in the following sections. The *hopalong split-and-merge* algorithm (Pavlidis 1982b) segments a planar curve into a minimal set of straight line segments while the *split-classify-merge-adjust* algorithm (Horaud and Bolles, 1983) segments a planar curve into a minimal set of straight line and curved segments.

Hopalong split-and-merge method: This segmentation algorithm as developed by Pavlidis (Pavlidis 1982b) provides a quick polygonal approximation of the stripe image contour as shown in Figure 3.9. This algorithm sequentially examines separate subsets of image points of specific size and splits and/or merges these subsets based on the results of colinearity and merging tests. The splitting occurs when the straight line joining the segment endpoints is not an accurate representation for all the

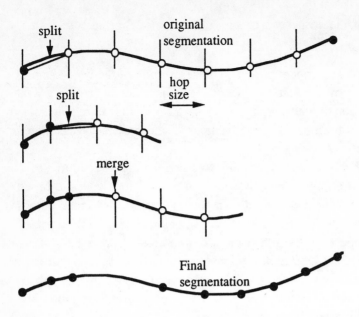

Figure 3.9 Hopskip algorithm for feature point detection.

points in the set. This is checked by measuring the maximum absolute error in the set. If this error is greater than the specified tolerance then the segment is split at the point farthest from the straight line and a new breakpoint inserted at this point. Two adjacent segments are merged if they satisfy the merging test which requires that the current and the previous segments be colinear. This is checked by measuring the distance of the common point of the two segments from the line joining the two endpoints of the merged segment. The angle between the two segments has also been used as a measure in some instances. If the distance to the common point is within a given tolerance, then the two segments can be merged into one. The value of the tolerance decides the accuracy of the representation but a tight tolerance drastically increases the number of segments required in the polygonal approximation of the feature surface (refer to Figure 3.10).

It should be noted that this algorithm completes the segmentation in one pass although the number of subsets will change during the processing as sets of points are split and/or merged.

Split-classify-merge-adjust algorithm: The steps in this algorithm are,

1.   split each unknown segment,

52

2. classify each new segment,
3. merge adjacent segments, and
4. adjust the breakpoints along the curve.

Steps (1) and (2) constitute the initial segmentation of the curve which is then followed by iterative segmentation, each iteration consisting of steps (3), (4), (1) and (2), in that order. The algorithm continues this iterative segmentation process until each segment is classified, well-fitted and no segment can be merged.

The initial segmentation in step (1) can be generated using Ramer's algorithm (Ramer 1972). This algorithm partitions a curved segment into straight segments by adding a breakpoint where the distance between the point and the chord joining the curve endpoints is greater than a specified tolerance. The amount of tolerance decides the accuracy of the representation. However, as mentioned earlier, it also decides the number of segments describing the curve. If the tolerance value is small then the number of segments increases drastically as shown in Figure 3.10. The threshold is selected corresponding to a point slightly to the right of the lower knee in Figure 3.10 (Horaud and Bolles, 1983).

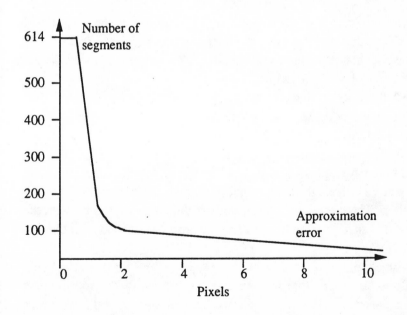

Figure 3.10 Number of linear segments for representing curves versus the allowable approximation error. (From Horaud and Bolles 1983).

53

The segment classification in step (2) is done by evaluating the fit between the segment points and the properties describing the three categories of segments: (1) line, (2) circular arc, and (3) unknown segment. Many different evaluation criteria have been proposed (Bolles and Fischler 1982; Pavlidis 1982b) and the initial classification can be performed on the basis of:

(1)   normalized area between the curve and the chord joining the end points (normalization is done by dividing the area by the square of the chord distance);

(2)   normalized maximum distance of the farthest point along the curve from the chord joining curve endpoints (normalization done by dividing by the chord length).

The initial classification procedure works well in identifying straight line segments. However, for circular arcs and undulatory segments, it may provide erroneous results. Hence, further evaluation is done to identify a segment as circular. This detailed procedure uses a sequence of errors to classify a segment as circular (Horaud and Bolles, 1983). It also identifies situations with effect of random noise. Those segments which fail are classified as unknown and subjected to yet another iteration of the algorithm.

Merging the adjacent segments in step (3) minimizes the number of segments describing the contour. The merge process is applicable only to segments which are classified as line or arc. The merging of two adjacent may result in a line or an arc as given below:

line + line   ->   line (colinear)
line + line   ->   arc
line + arc   ->   arc
arc + arc   ->   arc

The merger of two adjacent segments involves fitting an appropriate analytic curve through the merged segment and establishing its classification. The merger is finalized only if the error associated with the new segment is smaller than the larger of the errors associated with the two segments prior to merging.

The breakpoint adjustment in step (4) is a way of finetuning the overall algorithm since it provides a way of improving the approximate breakpoints inserted by the splitting step. The splitting algorithm positions a breakpoint $b(k)$ at a position where the distance between the curve and its chord is the farthest. The addition of a breakpoint results in two split segments ($s(k-1)$, $s(k)$) with their associated error between the curve and its chord. By repositioning the breakpoint $b'(k)$ within a specified range of the old breakpoint, the error associated with each split segment can be minimized. Mathematically, this can be modeled as

$$\| b'(k) - b(k) \| < range$$
$$\| e'(s(k-1), f) \| < \| e(s(k-1), f \|$$
$$\| e'(s(k), f) \| < \| e(s(k), f) \|$$

(3.2)

where,

　　　range is associated with the moving breakpoint,

　　　$e'(s(k-1), f)$ is the new error associated with segment s(k-1),

　　　$e'(s(k), f)$　is the new error associated with segment s(k),

　　　f is the analytic curve merging the two segments.

　　　The authors of this algorithm have established its convergence through reasoning that only segments classified as *unknown* are subjected to iterative segmentation. Hence at no step are the results of any previous steps undone to form an infinite loop. However, this algorithm does not guarantee convergence to a global minimum.

Refining Feature Points. As mentioned earlier, the identified feature points represent a coarse model of the stripe image. The coarse model contains effects of the limited camera resolution, quantization noise and other random noise. During seam tracking, the torch position and orientation is computed based on a model of the weld seam geometry, which in turn is computed from the feature point locations. Hence, for seam tracking application, refinement of the feature point locations is desirable to provide a more accurate model of the weld seam geometry. The process of refining the feature points involves:

1.　identifying the straight line segments (feature surfaces) in the stripe image bounded by the feature points;
2.　fitting a least square error line through the collection of these linear segments;
3.　computing the intersection points of adjacent least square error lines to find the new locations of the feature points.

The improvement in accuracy is attributed to the fact that while the original feature points are located on the stripe image and so affected by the camera's resolution, the refined location of the feature points can provide a higher resolution since more than two stripe image points are used in the computation.

3.4.2.3 Contour Feature Recognition. At this stage, specific feature recognition primitives can be used to extract and label specific feature points in the stripe image. For example, the left and right edge points can be located as the intersection of the two edges s_1 and s_2 (refer Figure 3.3 for vee-joint features). The root point in a vee-jointed weld preparation can be identified as the intersection of edges s_2 and s_3. For multipass situations or in the presence of tack welds along the seam, the wetting point can be located as the point on edge s_2 or s_3 which is closest to the tack weld.

3.5 SEAM ENVIRONMENT MODEL

The algorithms for feature extraction and seam type recognition associate a predefined seam type with the current scan. The inference about the seam type associated with each scan along with the location of the seam's characteristic features is used in modeling the seam environment. This model is essentially a database of the seam's history. A sample model generated after top-down feature extraction is shown in Table 3.1 for the unstructured seam environment illustrated in Figure 3.11.

The seam environment model is used to make realtime decisions regarding the torch-seam interaction. For example, the presence of a tack weld along the seam can be handled by an expert welding process controller

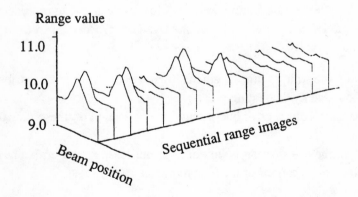

Figure 3.11 Sequence of range images representing an unstructured vee-jointed seam environment.

56

Table 3.1 Adaptive Model of the Unstructured Seam Environment.

Step k	Image No	Initial Decision about seam type	Decision Modified ?	Rules Used
0	0	Vee-joint (known input)	No	Rule: 21
1	1	Vee-joint	No	Rules: 03, 11, 21
2	2	Vee-joint	No	Rules: 03, 11, 21
3	3	Vee-joint	No	Rules: 03, 11, 21
4	4	Presently Unknown	Tack (at k = 6)	Rules: 01, 02, 13
5	5	Presently Unknown	Tack (at k = 6)	Rules: 01, 02, 11
6	6	Vee-joint	No	Rules: 03, 12, 21
7	7	Vee-joint	No	Rules: 03, 11, 21
8	8	Vee-joint	No	Rules: 03, 11, 21
9	9	Presently Unknown	End-of-seam	Rules: 01, 02, 13
10	10	Presently Unknown	End-of-seam	Rules: 01, 02, 11
11	11	Presently Unknown	End-of-seam	Rules: 01, 02, 11
12	12	End-of-seam	--	Rules: 01, 02, 11, 14

by either increasing the welding speed or increasing the arc current to burn through the tack weld. Additional functionality can be provided by the robotic welding system. through sophisticated seam environment models by including such comprehensive information as the seam's radius curvature, weld joint volumes, etc. Much of this information is computed from the features extracted in the earlier stages. Since aspects of welding process control are outside the scope of this monograph only the preliminary details regarding seam environment modeling are presented here.

3.6 SUMMARY

Feature extraction and seam recognition is crucial to intelligent realtime seam tracking. In an unstructured environment, although the various scenarios that the system would encounter during seam tracking are well defined, the actual sequence of their occurrence is not known *a priori*.

This chapter covered the details of processing both intensity and range images for feature extraction and seam recognition in an unstructured environment. To filter out the noise in both types of images, preprocessing using mathematical and heuristic methods is required. Seam recognition itself can be implemented using either of the two approaches: top-down method or the bottom-up method. The top-down approach starts with application of heuristics to hypothesize the most likely seam type in the current scan. This is followed by the application of feature extraction algorithms specific to that seam type and subsequent verification of the hypothesis based on the extracted features. Bottom-up approach on the other hand, specifies a weld joint as a collection of feature primitives such as straight lines and circular arcs, which are then recognized in the image. The objective of either approach is to associate a seam type with the current image and locate the various features in it. A history of this and other derived information forms a model of the seam environment that is used to control all torch-seam interaction.

For seam tracking systems, the robot motion control issues are equally important. The next chapter discusses the various coordinate frames involved in the computation of the input to the robot controller for positioning and orienting both the torch and the sensor along the seam. The methodology for defining these coordinate frames are also included in the discussion.

CHAPTER 4

COORDINATE FRAMES FOR REALTIME SEAM TRACKING

4.1 INTRODUCTION

Four coordinate frames play an important role in realtime path generation, viz.

1. Base coordinate frame ([W]);
2. Sensor coordinate frame ([S]);
3. Torchtip coordinate frame ([T]);
4. Seam coordinate frame ([C(k)]).

While the first three coordinate frames are fixed, either in space or to the robot arm, the Seam coordinate frame ([C(k)]) is associated with the seam and hence, varies at different locations along the seam. The relationship among the various coordinate frames is illustrated in Figure 4.1 and a detailed description of each coordinate frame is as follows.

4.2 BASE COORDINATE FRAME ([W])

This is the predefined coordinate frame associated with the robot arm and located at the base of the robot. For a single robot application, this coordinate frame is also usually the reference coordinate frame for the welding cell. In the case of a preprogrammed path seam tracking system, the position and orientation of the taught path can shift from part to part due to fixturing errors. For proper operation of such a system, the part positioning error when computed in the base coordinate frame can be corrected in all the original weld paths taught along the part. This needs to be done only once per part by offsetting the base coordinate frame itself corresponding to the part positioning error. This correction then affects all the original weld paths simultaneously.

In the case of non-preprogrammed seam tracking systems, the base coordinate frame provides a common reference for describing the torchtip, sensor, and the seam coordinate frames.

W: Base coordinate   frame

T: Torch-tip coordinate frame

S: Sensor coordinate frame

C(k-1): Seam coordinate frame at step (k-1).

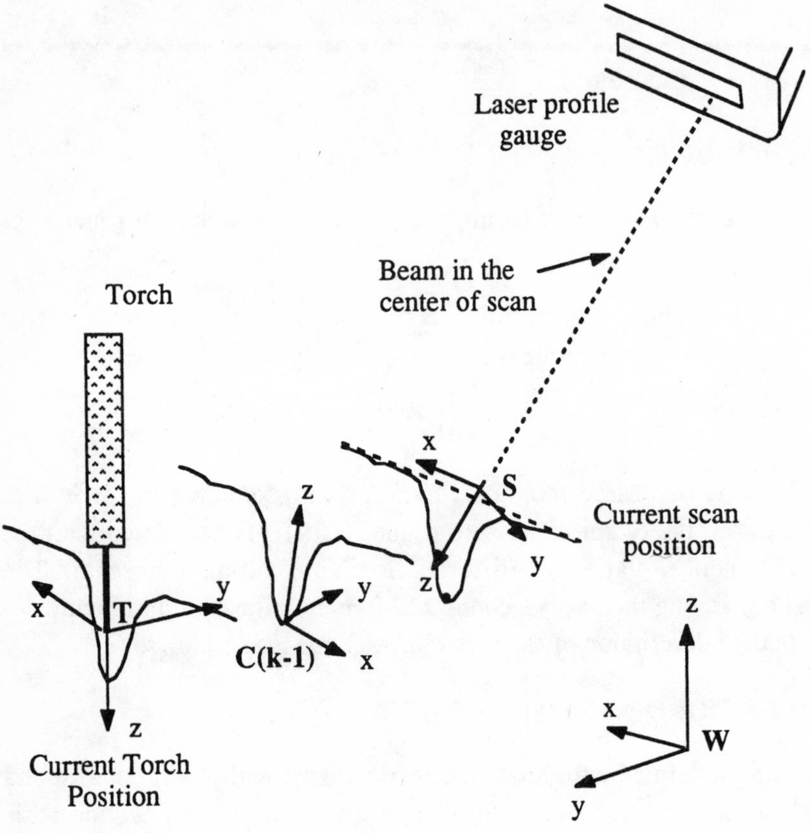

Figure 4.1 Coordinate frames for realtime seam tracking.

4.3 TORCHTIP COORDINATE FRAME [T]

This coordinate frame defines the position and orientation of the torchtool in the base coordinate frame. This coordinate frame with its origin at the torchtip is defined such that the z-axis is along the wirefeed direction and the x-axis and y-axis are defined arbitrarily under the constraint of righthand coordinate frame. During realtime seam tracking, the corrections to the

60

original trajectory are easiest implemented by applying them in the torchtip coordinate frame.

When the torch is not attached, the default tool is defined at the robot endeffector. Let this tool be denoted as $TOOL_0$. After mounting the torch at the robot endeffector, the Torchtip coordinate frame [T] is defined with respect to the default robot endeffector coordinate frame $TOOL_0$. Let this coordinate transformation be $TORCH\_TOOL$ with its two components: the translational component and the orientation component. The methodology for defining the tool at the torchtip is as follows.

4.3.1 Translation Component of $TORCH\_TOOL$

1. A pointer is rigidly mounted in the robot work envelope.
2. Setting $TOOL = TOOL_0$, the position of the pointer is recorded with no tool mounted at the endeffector, as shown in Figure 4.2. Let this location be NO_TOOLPT.
3. The torch is now mounted at the robot endeffector. Still keeping the $TOOL = TOOL_0$, the position of the pointer is recorded with the torchtip as shown in Figure 4.3. Let this location be $TOOLPT$.
4. The translational component of the coordinate transformation relating the robot endeffector coordinate frame and the Torchtip coordinate

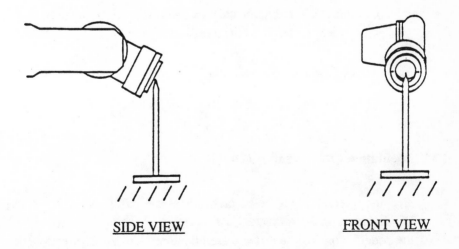

SIDE VIEW   FRONT VIEW

Figure 4.2 Learning the endeffector location for computing the translation component of the Torchtip coordinate frame [T].

61

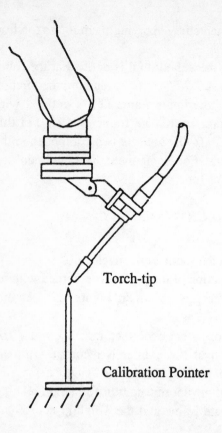

Torch-tip

Calibration Pointer

Figure 4.3 Learning the torchtip location for computing the translation
component of the Torchtip coordinate frame [T].

frame is obtained from the Equation (4.1)

$$[TOOL\_XYZ] = [TOOL\_PT]^{-1} [NOTOOL\_PT] \qquad (4.1)$$

4.3.2 Orientation Component of *TORCH_TOOL*

5.   Keeping the torchtip at the pointer, we set the *TOOL = TOOL_XYZ*.
Let this location be *NOROTPT*. At this stage, the coordinate frames at
the endeffector and the torchtip are oriented in the same direction, as
shown in Figure 4.4.

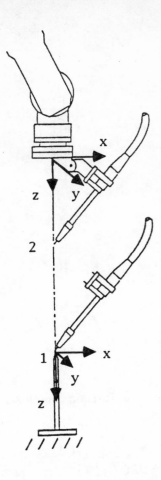

Figure 4.4 Physically marking the z-axis of the Torchtip coordinate frame.

6. The torchtip is now moved along the z-axis of the tool coordinate frame. The start and end points of the z-axis motion are physically marked by pointers to indicate the z-axis direction of *TOOL_XYZ* as illustrated in Figure 4.4.

7. With *TOOL = TOOL_XYZ*, the torch is now rotated about both, the x-axis and the y-*axis* to align the torch's wirefeed axis along the tool z-axis as illustrated in Figure 4.5. (the tool z-axis is obtained in step 6). Let this location be *ROTPT*.

8. The orientation of the Torchtip coordinate frame with respect to the robot endeffector coordinate frame is obtained from Equation (4.2),

63

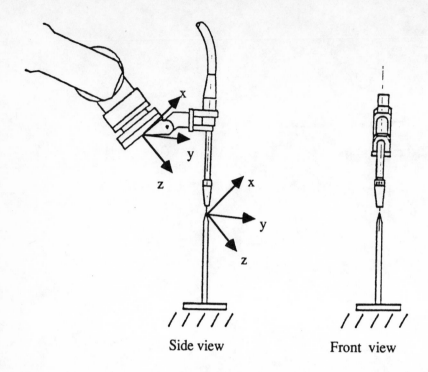

Figure 4.5 Learning the orientation component of the Torchtip coordinate frame.

$$[TOOL\_ROT] = [ROT\_PT]^{-1} [NOROT\_PT] \qquad (4.2)$$

9. The final coordinate transformation $TORCH\_TOOL$, relating the Torchtip coordinate frame and the robot endeffector coordinate frame is obtained by combining the translational and the orientation component obtained separately from Equations (4.1) and (4.2).

The above two step procedure establishes the Torchtip coordinate frame [T] at the torchtip with the z-axis along the wirefeed direction. This coordinate frame as defined in the Base coordinate frame [W] is given by Equation (4.3)

$$[\mathbf{T}] = [TOOL_0][TORCH\_TOOL] \qquad (4.3)$$

64

4.4 SENSOR COORDINATE FRAME [S]

The image coordinates are available in this coordinate frame. For a scanning beam range sensor, the image coordinates in the sensor coordinate frame are given by the position of the beam along the scan (p) and the associated range (r) to the spatial point being observed. For such a system, the sensor coordinate frame can be defined in the epipolar plane (scanning plane) of the range sensor defined by the beam source, target object, and the reflected beam into the camera. In the case of range sensors using an acousto-optic crystal to position the beam along the scan, this coordinate frame can be conveniently located in the center of the scanning window at a distance from the beam source. As shown in Figure 4.7 this distance is selected to be the range sensor's nominal standoff distance from the beam source. The *z-axis* in this coordinate frame is defined along the beam while the *x-axis* is along the direction of scan. The *y-axis* is determined by the right hand coordinate frame.

For a structured light projector-camera range sensor, the coordinates of the light stripe image are in the form of pixel locations (u, v) in the image space. For seam tracking, however, world coordinates of the joint geometry are required, which implies that the image coordinates have to be transformed into world coordinates. This transformation is computationally expensive and so is done only after most of the processing is done in the image coordinates. On the other hand, with scanning beam range sensors the resolution of the scan can be adjusted in order to work with a manageable data set. In any case, the transformation from the image space to the world space is an important step in computing the robot trajectory and is established through the process of *sensor calibration*.

4.4.1 Calibrating a Scanning Beam Range Sensor

For this kind of sensor, the range image consists of the beam position (p) along the scan and the associated range (r) to the spatial point (x, y, z) being observed. The first step is to map the 3D coordinates (x, y, z) of the observed point into the 2D coordinates (p, r) in image space using the principle of laser triangulation. The procedure of *camera calibration* establishes this mapping. The transformation of the 2D range image in sensor frame into the (x, y, z) world coordinates requires the sensor calibration procedure, which defines the sensor frame in world space. For seam tracking purposes, the coordinate transformation relating the sensor frame to the torchtip tool frame is also established. The procedures for camera calibration, Sensor coordinate frame [S] definition (sensor calibration), and the *Torch_to_Sensor* coordinate transformation [T_2S] definition for a scanning beam range sensor is as follows.

4.4.1.1 Camera Calibration. Camera calibration is required to map the 3D coordinates of a spatial point into the 2D coordinates of the range image (beam position and range). In the case of the scanning beam range sensor, the beam position is controllable during the scan and is therefore known. The camera calibration procedure establishes the range to the observed spatial point as a function of the beam position and the CCD element in linear array affected by the reflected beam (refer to Figure 4.6).

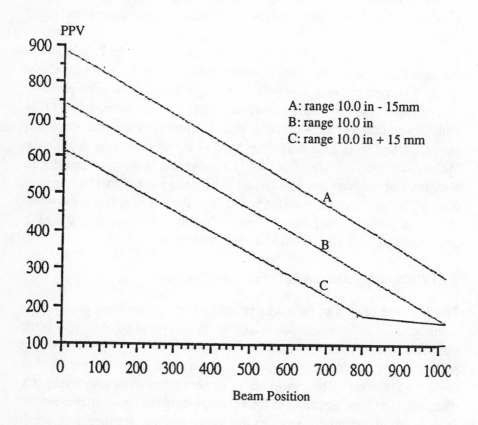

Figure 4.6 LPG Calibration data. PPV as a function of Range and beam position. (Courtesy of Chesapeake Laser Systems, Maryland, USA)

66

The camera calibration procedure can be implemented by setting a special planar target at known range values and then locating the CCD element affected for the set beam position. Figure 4.6 shows such a plot of the location of the affected CCD element against the beam position for a Laser Profiling Gage developed by the Chesapeake Laser Systems of Lanham, Maryland, USA. The carefully calibrated planar target was set at three known range values of 10.0 in–15 mm, 10.0 in, and 10.0 in+15 mm. As the beam was positioned along several known positions in the scan, the location of the affected CCD element in the camera was measured (described by an internal variable, ppv). A second order polynomial fitted through the ppv readings at each beam position (described by the internal variable, cco) gives the relation between the known beam position, the ppv reading, and the resulting range. A least squared error quadratic polynomial can be computed by positioning the planar target surface at more than three known range values.

The coefficients of this second order curve are stored for each position of the beam. During normal operation of the camera, the range to the observed spatial point at the known beam position can be computed by solving the corresponding second order polynomial in ppv values. Further details regarding this procedure are provided in Appendix A at the end of this chapter.

4.4.1.2 Defining the Sensor Frame ([S]). This coordinate frame is established through the process of sensor calibration and consists of two components:
1. coordinates of the sensor frame origin;
2. orientation of the sensor frame relative to the base frame.

Sensor Frame Origin Coordinates
1. Position the beam in the center of the scan as shown in Figure 4.7. In addition, move the sensor attached to the robot arm such that the range to the beam point on the plate is equal to the range sensor's nominal standoff distance.
2. Setting the *TOOL = TORCH_TOOL* (as explained in the previous section on Torchtip coordinate frame), record the location of the Torchtip coordinate frame [T]. Let this be *TORCH_POS*.
3. Physically mark the position of the beam point on the plate. This marks the origin of the Sensor coordinate frame [S] in the Base coordinate frame [W].

67

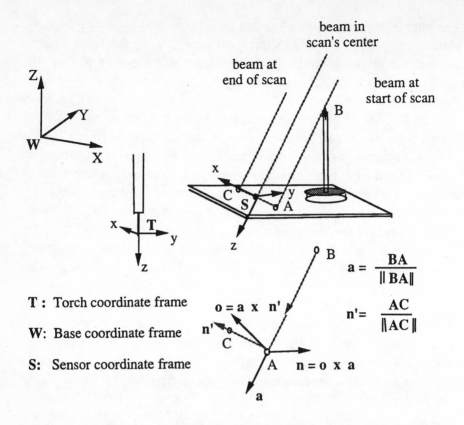

$$a = \frac{BA}{\|BA\|}$$

$$n' = \frac{AC}{\|AC\|}$$

$$o = a \times n'$$

$$n = o \times a$$

T : Torch coordinate frame

W: Base coordinate frame

S: Sensor coordinate frame

Figure 4.7 Defining the Sensor coordinate frame in the scanning beam
range sensor's epipolar plane.

4.   Still keeping the *TOOL = TORCH_TOOL*, move the torchtip to the
point marking the Sensor coordinate frame's origin and record its
location. Let this be *ORG* . This represents the sensor coordinate
frame's origin in the world space.

Sensor Frame Orientation
5.   Keeping the *TOOL = TORCH_TOOL*, move the torchtip back to
TORCH_POS. The beam should now be over the origin of the Sensor
coordinate frame, which is marked on the plate.
6.   Move the beam to the start of the scan. Mark this position of the beam
on the plate as **A**, as shown in Figure 4.7. Next, using a pointer, mark
another position **B** along the beam. The line **BA** represents the *z-axis*
direction of the Sensor coordinate frame.

7. Move the beam to the end of the scan. Mark this position of the beam on the plate as **C**.

8. Keeping $TOOL = TORCH\_TOOL$, move the torchtip to the marked points **A**, **B**, and **C** to record their locations in the Base coordinate frame.

9. Vector $\overline{BA}$ and vector $\overline{AC}$ will be orthogonal only if the view vector of the range sensor is normal to the plate. Hence, $\overline{AC}$ may not necessarily represent the *x*-axis direction. The unit vector **a**, along the Sensor coordinate frame's *z*-axis is given by

$$\mathbf{a} = \frac{\overline{BA}}{\left|\overline{BA}\right|} \tag{4.4}$$

The unit vector **n′** along $\overline{AC}$ is given by

$$\mathbf{n}' = \frac{\overline{AC}}{\left|\overline{AC}\right|} \tag{4.5}$$

Hence, the unit vector **o** along the Sensor coordinate frame's *y-axis* can now be found as a cross-product of **a** and **n′** and is given by

$$\mathbf{n}' = \frac{\mathbf{a} \times \mathbf{n}'}{\left|\mathbf{a} \times \mathbf{n}'\right|} \tag{4.6}$$

The unit vector **n** along the Sensor coordinate frame's *x*-axis is now given by

$$\mathbf{n} = \mathbf{o} \times \mathbf{a} \tag{4.7}$$

10. The Sensor coordinate frame [**S**], defined in the Base coordinate frame, is computed by combining the position coordinates *(ORG)* from step (4) and the orientation $\begin{bmatrix} \mathbf{n} & \mathbf{o} & \mathbf{a} \end{bmatrix}$ from step (9), respectively.

4.4.1.3 Torch_to_Sensor Coordinate Transformation [T2S]. Seam tracking application requires relationship between the sensor frame and the torchtip frame. This transformation is fixed since the torch and the range sensor are mounted at the robot endeffector and is given by Equation (4.8),

$$\mathbf{T_2 S} = \begin{bmatrix} TORCH\_POS \end{bmatrix}^{-1} \begin{bmatrix} S \end{bmatrix} \tag{4.8}$$

4.4.2 Calibrating a Projector-Camera Range Sensor

For this kind of sensor, sensor calibration allows extraction of 3D information from 2D images. The (x, y, z) coordinates of a point in space cannot be completely determined from its image (u, v) in pixel coordinates. This is due to the fact that the image coordinates represent all points along the ray of light rather than one single point. To remove this ambiguity, the sensor computes the 3D information about a point by establishing a one-to-one correspondence between points on the projected light plane and the

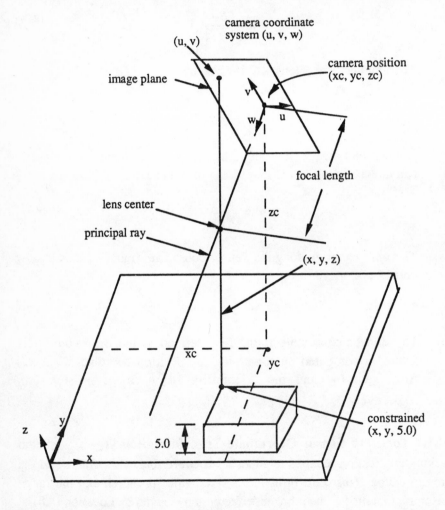

Figure 4.8 Camera geometry relating coordinates of a point in space to its image coordinates. (Courtesy of SRI International, California, USA)

70

image coordinates. Points in space where the light plane intersects the surface being measured, represent points which belong to both the light plane and the surface. The sensor calibration procedure establishes a relationship between the 2D image coordinates *(u, v)* and the 3D coordinates *(x, y, z)* of the surface point being measured, in either the world coordinate frame or the sensor coordinate frame.

Sensor calibration is a two step process that involves computing the camera matrix (representing camera's projective transformation mapping the 3D world coordinates into 2D image coordinates) and defining the light plane. The camera matrix and the light plane equations, given by Equations (4.9) and (4.12), can then be used to compute the sensor matrix which provides the *(x, y, z)* coordinates of a point in space based on its image coordinates *(u, v)* (Nitzan, et al. 1983).

Using homogeneous coordinates with scaling factor *s*, the 4 x 4 camera matrix transforms the *(x, y, z)* coordinates of a point in space into its image coordinates *(u, v)* (refer to Figure 4.8), given by equation (4.9).

$$
\begin{bmatrix} s \cdot u \\ s \cdot v \\ s \cdot w \\ s \end{bmatrix} = \begin{bmatrix} a_{11} & a_{12} & a_{13} & a_{14} \\ a_{21} & a_{22} & a_{23} & a_{24} \\ a_{31} & a_{32} & a_{33} & a_{34} \\ a_{41} & a_{42} & a_{43} & a_{44} \end{bmatrix} * \begin{bmatrix} x \\ y \\ z \\ 1 \end{bmatrix} \qquad (4.9)
$$

The image coordinates *(u, v)* represent a ray in space defined by the intersection of two planes *u* and *v* as illustrated in Figure 4.9 and given by

$$
u = \frac{a_{11} \cdot x + a_{12} \cdot y + a_{13} \cdot z + a_{14}}{a_{41} \cdot x + a_{42} \cdot y + a_{43} \cdot z + a_{44}} \qquad (4.10)
$$

$$
v = \frac{a_{21} \cdot x + a_{22} \cdot y + a_{23} \cdot z + a_{24}}{a_{41} \cdot x + a_{42} \cdot y + a_{43} \cdot z + a_{44}} \qquad (4.11)
$$

To uniquely determine the coordinates *(x, y, z)* of a point along the ray, we have to define a third plane which also contains the point. The equation of the third plane has to be known *a priori* such that when the two planes *u* and *v* are measured from the image coordinates, the 3D information *(x, y, z)* of the point is known. Let the equation of the third plane be

$$
b_1 \cdot x + b_2 \cdot y + b_3 \cdot z + b_4 = 0. \qquad (4.12)
$$

71

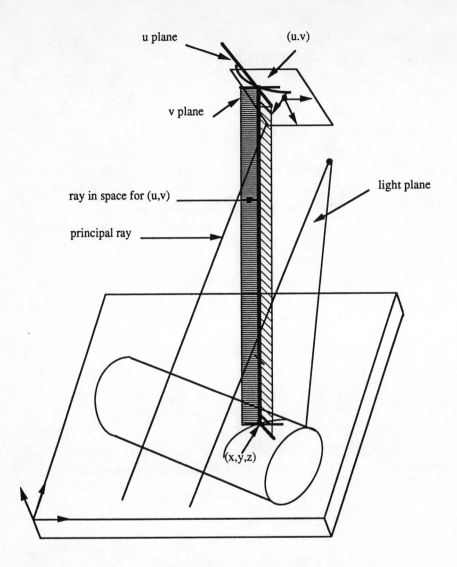

Figure 4.9 An *(x, y, z)* surface point defined by the intersection of the *u*, *v*, and light planes. (Courtesy of SRI International, California, USA)

Then the *(x, y, z)* coordinates of the point in space whose image is given by *(u, v)* is determined by solving the 3 equations

$$(a_{11} - u \cdot a_{41})x + (a_{12} - u \cdot a_{42})y + (a_{13} - u \cdot a_{43})z + \quad (4.13)$$
$$(a_{14} - u \cdot a_{44}) = 0$$

72

$$(a_{21} - v \cdot a_{41})x + (a_{22} - v \cdot a_{42})y + (a_{23} - v \cdot a_{43})z +$$

$$(a_{24} - v \cdot a_{44}) = 0 \qquad (4.14)$$

$$b_1 x + b_2 y + b_3 z + b_4 = 0 \qquad (4.15)$$

These equations in matrix form are

$$
\begin{bmatrix}
b_1 & b_2 & b_3 \\
(a_{11} - u \cdot a_{41}) & (a_{12} - u \cdot a_{42}) & (a_{13} - u \cdot a_{43}) \\
(a_{21} - v \cdot a_{41}) & (a_{22} - v \cdot a_{42}) & (a_{23} - v \cdot a_{43})
\end{bmatrix}
*
\begin{bmatrix}
x \\
y \\
z
\end{bmatrix}
=
\begin{bmatrix}
-b_4 \\
(u \cdot a_{44} - a_{14}) \\
(v \cdot a_{44} - a_{24})
\end{bmatrix}
$$

$$(4.16)$$

This matrix when inverted symbolically gives the sensor matrix (Nitzan et al. 1983), given by

$$
\begin{bmatrix}
s \cdot x \\
s \cdot y \\
s \cdot z \\
s
\end{bmatrix}
=
\begin{bmatrix}
m_{11} & m_{12} & m_{13} \\
m_{21} & m_{22} & m_{23} \\
m_{31} & m_{32} & m_{33} \\
m_{41} & m_{42} & m_{43}
\end{bmatrix}
*
\begin{bmatrix}
u \\
v \\
1
\end{bmatrix}
\qquad (4.17)
$$

where,

$$m_{11} = (b_4 \cdot a_{22} - b_2 \cdot a_{24}) \cdot a_{43} + (b_3 \cdot a_{24} - b_4 \cdot a_{23}) \cdot a_{42} + \qquad (4.18)$$
$$(b_2 \cdot a_{23} - b_3 \cdot a_{22})$$

$$m_{12} = (b_2 \cdot a_{14} - b_4 \cdot a_{12}) \cdot a_{43} + (b_4 \cdot a_{13} - b_3 \cdot a_{14}) \cdot a_{42} + \qquad (4.19)$$
$$(b_3 \cdot a_{12} - b_2 \cdot a_{13})$$

$$m_{13} = (b_2 \cdot a_{13} - b_3 \cdot a_{12}) \cdot a_{24} + (b_4 \cdot a_{12} - b_2 \cdot a_{14}) \cdot a_{23} + \qquad (4.20)$$
$$(b_3 \cdot a_{14} - b_4 \cdot a_{13}) \cdot a_{22}$$

$$m_{21} = (b_1 \cdot a_{24} - b_4 \cdot a_{21}) \cdot a_{43} + (b_4 \cdot a_{23} - b_3 \cdot a_{24}) \cdot a_{41} + \qquad (4.21)$$
$$(b_3 \cdot a_{21} - b_1 \cdot a_{23})$$

$$m_{22} = (b_4 \cdot a_{11} - b_1 \cdot a_{14}) \cdot a_{43} + (b_3 \cdot a_{14} - b_4 \cdot a_{13}) \cdot a_{41} + \qquad (4.22)$$
$$(b_1 \cdot a_{13} - b_3 \cdot a_{11})$$

$$m_{23} = (b_3 \cdot a_{11} - b_1 \cdot a_{13}) \cdot a_{24} + (b_1 \cdot a_{14} - b_4 \cdot a_{11}) \cdot a_{23} + \qquad (4.23)$$
$$(b_4 \cdot a_{13} - b_3 \cdot a_{14}) \cdot a_{21}$$

$$m_{31} = (b_4 \cdot a_{21} - b_1 \cdot a_{24}) \cdot a_{42} + (b_2 \cdot a_{24} - b_4 \cdot a_{22}) \cdot a_{41} + \qquad (4.24)$$
$$(b_1 \cdot a_{22} - b_2 \cdot a_{21})$$

$$m_{32} = (b_1 \cdot a_{14} - b_4 \cdot a_{11}) \cdot a_{42} + (b_4 \cdot a_{12} - b_2 \cdot a_{14}) \cdot a_{41} + \qquad (4.25)$$
$$(b_2 \cdot a_{11} - b_1 \cdot a_{12})$$

$$m_{33} = (b_1 \cdot a_{12} - b_2 \cdot a_{11}) \cdot a_{24} + (b_4 \cdot a_{11} - b_1 \cdot a_{14}) \cdot a_{22} + \qquad (4.26)$$
$$(b_2 \cdot a_{14} - b_4 \cdot a_{12}) \cdot a_{21}$$

$$m_{41} = (b_2 \cdot a_{21} - b_1 \cdot a_{22}) \cdot a_{43} + (b_1 \cdot a_{23} - b_3 \cdot a_{21}) \cdot a_{42} + \qquad (4.27)$$
$$(b_3 \cdot a_{22} - b_2 \cdot a_{23}) \cdot a_{41}$$

$$m_{42} = (b_1 \cdot a_{12} - b_2 \cdot a_{11}) \cdot a_{43} + (b_3 \cdot a_{11} - b_1 \cdot a_{13}) \cdot a_{42} + \qquad (4.28)$$
$$(b_2 \cdot a_{13} - b_3 \cdot a_{12}) \cdot a_{41}$$

$$m_{43} = (b_2 \cdot a_{11} - b_1 \cdot a_{12}) \cdot a_{23} + (b_1 \cdot a_{13} - b_3 \cdot a_{11}) \cdot a_{22} + \qquad (4.29)$$
$$(b_3 \cdot a_{12} - b_2 \cdot a_{13}) \cdot a_{21}$$

The sensor matrix coefficients m_{ij} in equation (4.17) have to be determined, which in turn require the camera matrix coefficients a_{ij} and the light plane coefficients b_i. This is done using the process of camera calibration and light plane calibration, which are discussed below.

4.4.2.1 Camera Calibration. The sensor matrix is derived from elements of the camera matrix as given by equations (4.17)–(4.29). The camera matrix can be computed directly from the image coordinates corresponding to known world points. To implement this scheme, a specially made calibration target is set at known positions in the camera's field of view. Then a set of linear equations is formed from the known world locations of the target points and their corresponding image positions. These linear equations can be of the form shown in Equations (4.13) and (4.14). However, further simplification is possible if one represents the camera matrix in the form

$$
\begin{bmatrix} s \cdot u \\ s \cdot v \\ s \cdot w \\ s \end{bmatrix} = \begin{bmatrix} a_{11} & a_{12} & a_{13} & a_{14} \\ a_{21} & a_{22} & a_{23} & a_{24} \\ 0 & 0 & 1 & 0 \\ a_{41} & a_{42} & a_{43} & a_{44} \end{bmatrix} * \begin{bmatrix} x \\ y \\ z \\ 1 \end{bmatrix} \tag{4.30}
$$

This form is possible for two reasons: First, since the 2D image coordinates represent a ray of light, the value of the third coordinate (w) can be arbitrarily set. Second, since the value of the scale factor is explicitly included in the homogeneous transformation, the camera matrix can be multiplied by any factor throughout without affecting the transformation. Thus we can choose a factor such that the element $a_{44} = 1$ in the modified transformation. Equations (4.13) and (4.14) with $a_{44} = 1$ to give

$$
u = a_{11} \cdot x + a_{12} \cdot y + a_{13} \cdot z + a_{14} - \tag{4.31}
$$
$$
u \cdot a_{41} \cdot x - u \cdot a_{42} \cdot y - u \cdot a_{43} \cdot z
$$

$$
v = a_{21} \cdot x + a_{22} \cdot y + a_{23} \cdot z + a_{24} - \tag{4.32}
$$
$$
v \cdot a_{41} \cdot x - v \cdot a_{42} \cdot y - v \cdot a_{43} \cdot z
$$

Thus for each pair of known world points (x, y, z) and their corresponding image coordinates (u, v), we get two equations described by Equations (4.31) and (4.32). For an experiment with N pairs, we get 2N constraints in the form of 2N x 11 matrix. The resulting least squares estimates of the camera matrix elements can be solved using techniques for solving a set of overconstrained linear equations. This completes the camera calibration process.

4.4.2.2 Calibration of the light plane. This process determines the equation of the light plane as given by Equation (4.15) and required for computing the sensor matrix. The equation of the light plane is determined by fitting a plane through a set of points known to lie in the light plane. This can be implemented by intersecting the light plane with a planar object of known world z-coordinates and measuring the (x, y, z) coordinates using the camera

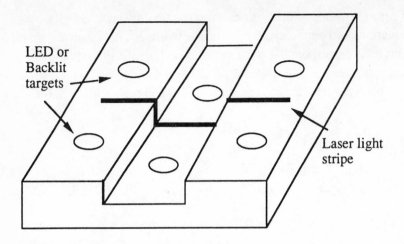

LED or Backlit targets

Laser light stripe

Figure 4.10. Fixture for calibrating the light plane.[†]

matrix defined earlier. A set of points can thus be measured by either changing the z-coordinate position of the planar object while keeping the camera stationary or vice versa if the camera is mounted on an arm. Another approach is to make a specially calibrated target object with planes at different z-levels as shown in Figure (4.10).

Remark: The relationship between the camera and the light source in a projector-camera range sensor is critical to its operation. If the distance between the camera and the projector (baseline) is too large, then an increased shadow region (points in space which are not simultaneously visible from the camera and the projector) can make the sensor ineffective. On the other hand, reducing the baseline reduces the shadow region but also reduces the resolution of the range sensor (a large change in the range is required to cause a change in the image coordinates). By increasing the focal length of the camera lens, the resolution can be improved, however, this reduces the field of view of the camera. Thus, the design of any range sensor

[†] Reprinted from Int. Journal of Robotics Research, Vol. 9:5, 1990, Agapakis et al., Approaches for recognition and interpretation of workpiece surface features using structured lighting, by permission of the MIT Press, Cambridge, Massachusetts, Copyright 1990 Massachusetts Institute of Technology.

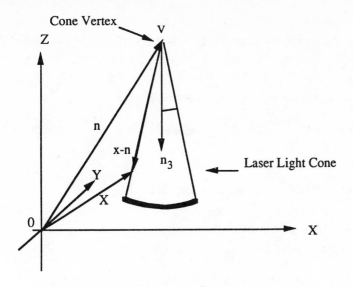

Figure 4.11. Calibration of a light cone.[†]

based on the triangulation principle is a tradeoff regarding the shadow area, the resolution and the field of view. ♣

4.4.2.3 Calibration of a Light Cone.

Certain seam tracking systems prefer the use of a light cone rather than a light plane to illuminate the surface of the seam. The tracking system developed by Automatix Inc., Billerica, Massachusetts is one such example (Agapakis, 1990b). According to the authors , the cone of light has certain distinct advantages over the light plane approach. The image processing can now be done in a radial window making it is easier to remove noise due to welding spatter which is expected to have a radial trajectory when flying out from the torch.

A system using a light cone is very similar to a system using a light plane when it comes to defining the camera matrix and the sensor matrix (refer Figure 4.11). The difference lies in the calibration of the light cone.

[†] Reprinted from Int. Journal of Robotics Research, Vol. 9:5, 1990, Agapakis et al., Vision-aided robotic welding: an approach and a flexible implementation, by permission of the MIT Press, Cambridge, Massachusetts, Copyright Massachusetts Institute of Technology 1990.

This can be determined by fitting a cone through a set of points known to lie on the cone. This can be implemented by intersecting the light cone with a planar surface of known world z-coordinates and then measuring the *(x, y, z)* coordinates using the camera matrix defined earlier. The light cone in 3D space can be represented in quadratic form as

$$\left| \mathbf{x} - \mathbf{n} \right|^2 = \sec^2 \zeta \left[(\mathbf{x} - \mathbf{n}) \bullet \mathbf{n}_3 \right]^2 \tag{4.33}$$

where,

$\mathbf{x}$:   is a vector to any point on the cone surface $\mathbf{x} = \begin{bmatrix} x, & y, & z \end{bmatrix}^T$;

$\mathbf{n}_3$:   is the unit vector along the cone axis;

$\mathbf{n}$:   is the vector to the cone vertex;

ζ: is the cone vertex half angle.

The cone parameter (semiangle of the cone) is given by

$$\cos \zeta = \frac{\left[(\mathbf{x} - \mathbf{n}) \bullet \mathbf{n}_3 \right]}{\left| \mathbf{x} - \mathbf{n} \right|} \tag{4.34}$$

Representing in quadratic form, the cone can be described in 3D as given by Equation (4.33). This is a special case of the generalized representation of any conic section in 3D space which is given by

$$\mathbf{x}^T \mathbf{A} \mathbf{x} + \mathbf{b}^T \mathbf{x} + c = 0 \tag{4.35}$$

where,

$\mathbf{A}$:   is a symmetric 3 x 3 matrix;

$\mathbf{b}$:   is a 3 element vector;

c: is a scalar constant.

By comparing Equations (4.33) and (4.35), the following relationships can be established

$$\mathbf{A} = \mathbf{I} - \sec^2 \zeta \, \mathbf{n}_3 \, \mathbf{n}_3^T$$

$$\mathbf{b} = -2 \mathbf{A} \mathbf{n}$$

$$c = \mathbf{n}^T \mathbf{A} \mathbf{n}$$

where $\mathbf{I}$ is an identity matrix.

After normalization, the Equation (4.35) has nine independent parameters and can be applied to any conic section. However, if we use the

78

special case of a cone, then we can use Equation (4.33) which has only six independent parameters.

The calibration of the light cone is implemented by illuminating a planar calibration fixture from different heights and computing the 3D coordinates of the laser stripe from its 2D image coordinates. By measuring the (x, y, z) coordinates at many points, one can use the linear least squares fit to obtain the independent parameters of the generalized equation (4.35). In the special case of a cone, it is possible to linearize the quadratic form around the nominal parameters of the cone known *a priori* from hardware specifications. Linear least squares can then be employed to compute the deviations from the nominal parameters of the cone.

For the case of intersection between a light cone and a planar surface, given the image coordinates $\mathbf{u} = [u, v]^T$, the camera matrix $[\mathbf{C}, \mathbf{d}]$, and the cone coefficients $[\mathbf{A}, \mathbf{n}]$, by eliminating $\mathbf{x} = \begin{bmatrix} x, & y, & z \end{bmatrix}^T$ from Equations (4.9) and (4.33) we get a quadratic equation in the scaling factor s,

$$s^2 \mathbf{u}^T \alpha \, \mathbf{u} + s \, \beta^T \mathbf{u} + \gamma = 0, \qquad (4.36)$$

where,

$$\alpha = (\mathbf{C}^{-1})^T \mathbf{A} (\mathbf{C}^{-1})$$
$$\beta = -2 (\mathbf{C}^{-1})^T \mathbf{A} (\mathbf{C}^{-1}\mathbf{d} + \mathbf{n})$$
$$\gamma = \mathbf{d}^T (\mathbf{C}^{-1})^T \mathbf{A} \mathbf{C}^{-1}\mathbf{d} + 2\mathbf{n}^T \mathbf{A} \mathbf{C}^{-1}\mathbf{d} + \mathbf{n}^T \mathbf{A} \mathbf{n}$$

Note: The camera matrix $[\mathbf{C}, \mathbf{d}]$, is computed earlier (refer Equation 4.9) and is given by

$$[\mathbf{C}, \quad \mathbf{d}] = \begin{bmatrix} a_{11} & a_{12} & a_{13} & a_{14} \\ a_{21} & a_{22} & a_{23} & a_{24} \\ a_{31} & a_{32} & a_{33} & a_{34} \\ a_{41} & a_{42} & a_{43} & a_{44} \end{bmatrix}$$

The correct root of s is selected based on the location of the camera pinhole whether it is inside or outside the light cone and based on which sector of the cone corresponds to the physical laser stripe. After solving equation (4.36) for s, one can substitute s and $\mathbf{u} = [u, v]^T$ in equation (4.9) to solve for $\mathbf{x} = \begin{bmatrix} x, & y, & z \end{bmatrix}^T$.

4.5 SEAM COORDINATE FRAME [C(K)]

This is a moving coordinate frame defined along the seam, with its origin on the root curve. The *y-axis* of this coordinate frame represents the seam direction while the z-axis is oriented normal to the seam. The x-axis is determined according to the right-handed coordinate system. This coordinate frame is defined in the cell's reference coordinate frame, which is also the Base coordinate frame. The computation of this coordinate frame from the range image is discussed later in Section 5.4.2.1.

4.6 SUMMARY

Realtime seam tracking requires the use of four coordinate frames, viz., Base coordinate frame [W], Torchtip coordinate frame [T], Sensor coordinate frame [S], and the Seam coordinate frame [C(k)]. For seam tracking application, the image coordinates in sensor space have to be transformed into world space for computing welding torch trajectory. This is done through the process of sensor calibration which allows computing the 3D coordinates *(x, y, z)* of an observed spatial point from its 2D image coordinates. Sensor calibration for a scanning beam range sensor is a two step process–camera calibration and establishing a sensor coordinate frame in the sensor's epipolar plane. For a projector-camera system, sensor calibration involves computing a sensor matrix from the camera matrix and the light plane coefficients.

Having discussed the coordinate frames for realtime seam tracking, the next chapter discusses the theory of seam tracking in an unstructured environment. Further details on the computing the Seam coordinate frame and its significance are also presented in greater details.

APPENDIX A

A.1 PPV TO RANGE CONVERSION

This appendix presents the camera calibration details for the Laser Profiling Gage (LPG) developed by Chesapeake Laser Systems, Lanham, Maryland, USA. Based on the principle of laser triangulation, the range to an observed spatial point is a function of (i) position of the beam in the scan, and (ii) the CCD element in the linear array affected by the reflected beam (outputted by the LPG as the preprocessor value called *ppv*). The process of camera calibration establishes a relationship between the ppv value and the beam position for a set of known ranges.

Figure A.1 shows the graph of *ppv* readings for different beam positions in

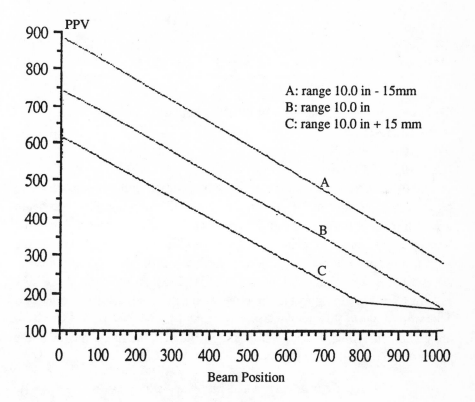

Figure A.1 LPG Calibration data. PPV as a function of Range and beam position. (Courtesy of Chesapeake Laser Systems, Maryland, USA)

Table A.1 CCO values for positioning the beam in a scan [0,800]

Beam Position	CCO	Beam Position	CCO
0	5660	450	10416
50	6021	500	10832
100	6501	550	12910
150	7503	600	13052
200	9157	650	13190
250	9339	700	13343
300	9553	750	13508
350	9795	800	13681
400	10079		

the interval [0, 1001], obtained for the three ranges:
(A)   10 in.–15 mm.
(B)   10 in.
(C)   10 in. + 15 mm.

In the graph each increment of the beam position (independent variable) represents a distance of 26 microns. In graph C, we notice that the beam goes out of sight for beam positions greater than 800 and therefore, only the beam positions in the interval [0,800] are useful for the entire dynamic range $\pm$ 15 mm. For each beam position in the interval [0,800], a least squares, second order approximating polynomial is fitted through the three *ppv* readings corresponding to the three preset range values mentioned earlier. These three sets of polynomial coefficients, i.e., a_{i0}, a_{i1}, and a_{i2}, $i = 0 \ldots 800$, are computed offline. For a given beam position (i), the range relates to the *ppv* through the following equation.

$$range = a_{i0} + a_{i1} \cdot ppv + a_{i2} \cdot ppv^2 \qquad \text{(A.1)}$$

During the LPG operation, the beam position is controllable and the ppv output is used in computing the range using Equation (A.1).

Table A.2 Coefficients of the 2nd order approximating polynomial for ppv.

Beam Position	a_0	a_1	a_2
0	1.457415E+1	-7.995831E-3	2.422492E-6
50	1.406400E+1	-7.003149E-3	1.819229E-6
100	1.382031E+1	-6.733723E-3	1.679061E-6
150	1.373768E+1	-6.929092E-3	1.883187E-6
200	1.359449E+1	-6.954763E-3	1.960274E-6
250	1.342013E+1	-6.865203E-3	1.941645E-6
300	1.318270E+1	-6.569204E-3	1.760544E-6
350	1.313116E+1	-6.928926E-3	2.140911E-6
400	1.292607E+1	-6.743382E-3	2.070237E-6
450	1.274252E+1	-6.608831E-3	1.992143E-6
500	1.256484E+1	-6.515041E-3	1.997734E-6
550	1.248363E+1	-6.819321E-3	2.409465E-6
600	1.231273E+1	-6.716709E-3	2.364853E-6
650	1.212369E+1	-6.572999E-3	2.322997E-6
700	1.192211E+1	-6.303034E-3	2.055811E-6
750	1.176286E+1	-6.288938E-3	2.182074E-6
800	1.163051E+1	-6.405119E-3	2.461614E-6

Remark: For experimental purposes, every 50th beam position is used in the interval [0, 800], thus giving a range with 17 points. The CCO values used for deflecting the beam to these positions are given in Table A.1. The polynomial coefficients used for computing the range values in this scan are given in Table A.2.   ♣

CHAPTER 5

SEAM TRACKING IN AN UNSTRUCTURED ENVIRONMENT: A CASE STUDY WITH VEE-GROOVE JOINTS

5.1 OVERVIEW

This chapter discusses the algorithms developed for a seam tracking system comprising of a six-axis, articulated joint robot and a scanning beam range sensor. Operation of an adaptive robotic welding system within the unstructured environment of a general 3D seam involves, (1) controlling the torch-seam interaction for achieving the desired weld qualities, and (2) controlling the position and orientation of the torch as well as the rotation of the lookahead range sensor for tracking the seam. These functions are governed by the high-level controller and the low-level controller, respectively. The inputs to these controllers are derived from the analysis of the range images. Specifically, the range image processing stage extracts the joint features and outputs their coordinates in world space, which is then used in the generation of the seam environment model and the seam geometry model.

Discussed here in detail are the algorithms for range image processing, seam environment modeling, seam geometry modeling, and the torch path generation for seam tracking in an unstructured environment. The results of the various algorithms are presented in the context of tracking a vee-groove on an aluminum plate, approximately one inch thick. Although the basic range image processing algorithm for vee-grooves was presented in Chapter 3, enhancements to the algorithm for operation in an unstructured environment are discussed in this chapter. The high-level control strategy, which primarily affects the welding process control, is presented in only so much detail as to stir the reader's imagination.

The major focus of this chapter is on the low-level control functions including, range image processing, modeling the seam geometry in terms of both, a continuous model and a discrete model, and the subsequent robot motion control for seam tracking. The path generation algorithm computes the position and orientation of the torch along the seam (5 DOF) and the rotation of the range sensor about the torch (1 DOF) for viewing the seam. These pathpoints, in cartesian frame, are input to the robot controller, which computes the joint angles using the inverse kinematic solution of the robot arm. The joint motion is controlled by the joint servo-controller. The joint angle computation and joint servo control is discussed elsewhere and so is not covered in this monograph. The output of the pathpoint generation algorithm can be input to most commercially available robot controllers for implementing seam tracking.

5.2 DEFINITIONS

$[\mathbf{R}(k)]$: (4 x 4) homogeneous transformation matrix in the Denavit-Hartenberg form (Denavit and Hartenburg 1955), describing the location of the Torchtip coordinate frame relative to the Base coordinate frame of the robot, during the (k)th seam tracking cycle. This coordinate frame is computed by the robot controller from the arm model and joint encoder values recorded at the start of the kth step.

Remark: The lookahead range sensor is scanning the seam n steps ahead of the current location of the torch tip. Hence $[\mathbf{R}(k)]$ represents the torch location during the (k+n)th scan by the scanning beam range sensor. ♣

$[\mathbf{R_I}]$: (4 x 4) homogeneous transformation matrix describing the initial location of the torch at the seam start, relative to the Base coordinate frame. This location is decided by the operator or programmed from a CAD database.

$[\mathbf{T_2 S}]$: (4 x 4) Fixed coordinate transformation describing the Sensor coordinate frame $[\mathbf{S}]$ relative to the Torchtip coordinate frame $[\mathbf{T}]$. This coordinate transformation remains constant during seam tracking because the configuration of the torch-sensor assembly mounted on the robot endeffector does not change. However, this coordinate transformation should be recalibrated in the event of a torch collision, etc. The method for obtaining this coordinate transformation is described in Chapter 4.

$[\mathbf{R}(-n)]$: (4 x 4) Transformation matrix describing the location of the torch, in the Base coordinate frame, when the range sensor's beam is positioned at the start of the seam for the 0th scan and is given by

85

$$[\mathbf{R}(-n)] = [\mathbf{R}_I][\mathbf{T}_2\mathbf{S}]^{-1} \qquad (5.1)$$

Remark: Tracking the seam from its beginning requires the beam to be positioned at the start of the seam. To reach this position of the range sensor, the torch is moved back from its location at the seam's start $[\mathbf{R}_I]$ to a location $[\mathbf{R}(-n)]$, which is outside the seam. This motion brings the Sensor coordinate frame to the start of the seam, where the Torchtip coordinate frame was originally located, in order to start the scanning process.   ♣

$[\mathbf{M}(k)]$: (4 x 4) homogenous transformation that describes the Interpretation coordinate frame relative to the Base coordinate frame of the robot. The frequently changing relationship between the seam and the lookahead range sensor may not allow a desirable view vector for the range sensor. The kth Interpretation coordinate frame, as the name suggests, is used in interpreting the kth range image before applying the image segmentation (feature extraction) algorithms discussed in Chapter 3. The Interpretation coordinate frame is given by

$$[\mathbf{M}(k)] = [\mathbf{C}(k-1)] \qquad (5.2)$$

where $[\mathbf{C}(k-1)]$ is the $(k-1)^{th}$ Seam coordinate frame.

$[\mathbf{R}_2\mathbf{C}]$: (4 x 4) Transformation matrix describing a rotation of 180 degrees about the x-axis of the Torchtip coordinate frame.

$[\mathbf{C}_2\mathbf{R}]$: (4 x 4) Transformation matrix describing a rotation of 180 degrees about the x-axis of the Seam coordinate frame.

$[\mathbf{D}(l)]$: (4 x 4) Drive transformation for linearly interpolating a coordinate frame between any two coordinate frames, $[\mathbf{R}(i)]$ and $[\mathbf{R}(j)]$ (Paul 1981). The drive transformation is a function of the relative motion parameter l where $0 \le l \le 1$.

Remark: As is discussed in later sections, the drive transformation is primarily used for calculating equidistant torch locations, which lie in between two measured (known) coordinate frames $[\mathbf{R}(i)]$ and $[\mathbf{R}(j)]$ whose relationship varies with i and j. The relative motion parameter l represents the relative distance of the interpolated location from the ith coordinate frame. Therefore,

$$\left[\mathbf{R}_{\text{interpolated}}\right] = [\mathbf{R}(i)][\mathbf{D}(l)] \qquad (5.3)$$

while

$$[\mathbf{R}(i)] = [\mathbf{R}(i)][\mathbf{D}(0)] \qquad (5.4)$$
$$[\mathbf{R}(j)] = [\mathbf{R}(i)][\mathbf{D}(1)]$$

♣

$\{S_i(k)\}$: Range image obtained from the kth scan consisting of r range data values in the Sensor coordinate frame (image space). The data $S_i(k)$ at each of the r points along the range image is given by a 4-element vector as follows.

$$S_i(k) = \begin{bmatrix} x_i = id_b \\ y_i = 0.0 \\ z_i \\ s = 1.0 \end{bmatrix} \qquad \text{for} \quad i = 0...r-1, \qquad (5.5)$$

where,

    d_b: spacing between adjacent scan points through which the beam is moved by the scanning beam range sensor;

    x_i: position of the beam at the ith scan point in image space;

    y_i: =0.0 since the scan is in sensor frame's xz-plane;

    z_i: range to the observed surface at the ith scan point;

    s: scaling factor.

$\{P_i(k)\}$: Range image after transformation from the Sensor coordinate frame into the kth Interpretation coordinate frame.

$\{P'_i(k)\}$: k^{th} range image in Interpretation coordinate frame after filtering and bad data suppression.

$\left\{ \mathbf{Q}_j(k) \right\}$: Feature point coordinates of the vee-joint in the Sensor frame. Since a vee-joint has three feature points, associated with the root and the two edges, $j = 1 \dots 3$.

$\left\{ \mathbf{w}_j(k) \right\}$: Feature point coordinates in world space (workcell frame). With a single robot in the workcell, the workcell coordinate frame is the same as the Base coordinate frame of the robot.

5.3 RANGE IMAGE PROCESSING

The algorithms for processing images of a vee-groove (top-down approach and bottom-up approach) are presented in Chapter 3. However, the top-down approach is further enhanced here to reliably perform in an unstructured environment. Range image interpretation is difficult in the Sensor coordinate frame because the view vector may not always be normal to the seam, resulting in nonsegmentable range images This difficulty arises due to the sensor having only 1 DOF, i.e., rotation about the torch axis, which changes the sensor view vector from scan to scan. The reliability of range image processing is improved by first transforming the image into the Interpretation Frame, and subsequently applying the top-down feature extraction algorithm as discussed in Chapter 3.

The feature points of a vee-groove include the root point and the joint edge points. Figures 5.1(a) and 5.1(b) illustrate the feature points as seen in the profile as well as its corresponding range image. Upon extraction, these

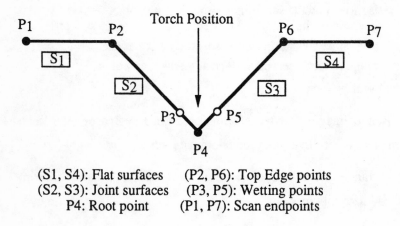

(S1, S4): Flat surfaces   (P2, P6): Top Edge points
(S2, S3): Joint surfaces   (P3, P5): Wetting points
P4: Root point   (P1, P7): Scan endpoints

Figure 5.1(a) Characteristic features of a vee-joint.

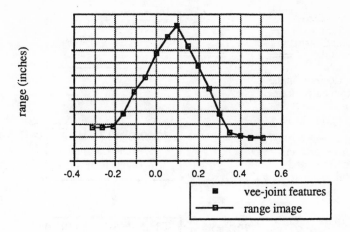

Figure 5.1(b) Vee-joint features as seen in the range image.

feature points are transformed into the Base frame and subsequently used in various seam models.

Figure 5.2 illustrates the flowchart of the Range Image Processing algorithm. In the flowchart, the blocks represent operations within the range image processing algorithm along with the appropriate input and output. The terminology used is consistent with that defined in Section 5.2. The step by step working of the algorithm is explained below.

5.3.1 Data Acquisition

In the sensory data acquisition task, two different sensors are involved, namely, the joint encoders in the robot arm and the range sensor. At the beginning of the kth seam tracking control cycle, the following information is acquired for computing the world coordinates of the seam's feature points:

1. After scanning the seam, the kth range image $\{\mathbf{S}_i(k)\}$ consisting of r range values in the Sensor frame. The range image acquired can be expected to be noisy due to effects of multiple reflections within the vee-groove. The image is filtered and smoothed using techniques described in Chapter 3.

Remark: In our experimental setup, the beam could be positioned at 1000 different locations in the scan, adjacent locations being 26 microns apart.

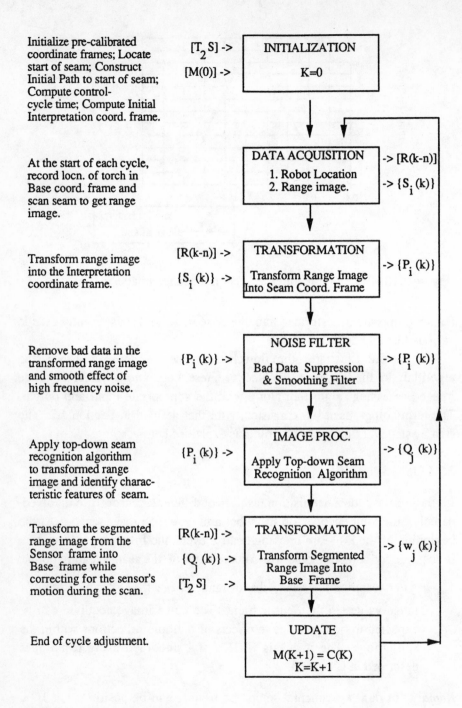

Figure 5.2 Flowchart of the range image processing algorithm.

90

For our experiments, we selected a scan approximately 0.8 inches wide with the beam positioned at 17 scanpoints ($r = 17$), each 1300 microns apart. ♣

2. The location of the Torchtip coordinate frame $[\mathbf{R}(k-n)]$. This coordinate frame is computed by the robot controller from the robot arm model and the joint encoder values recorded at the start of the kth scan.

Remark: Using the robot's internal joint encoders is one way to locate the endeffector in world space. This method, although very straightforward, has its drawbacks when high accuracy is required. The alternate method of using external locating devices such as laser interferometer, etc. provide high accuracy but are much more expensive. Further details on this subject are discussed in Appendix D at the end of Chapter 7. ♣

5.3.2 Range Image Transformation

A prerequisite for applying the top-down feature extraction algorithm as described in Chapter 3 is that range images should be segmentable. For a distorted range image $\{\mathbf{S}_i(k)\}$ in the Sensor frame, (possibly due to the varying relationship between the range sensor and the seam), feature extraction is made possible by first transforming it into the Interpretation coordinate frame $[\mathbf{M}(k)]$. The range images, before and after the transformation, are shown in Figures 5.3(a,b) through 5.7(a,b). These figures also illustrate the results of applying the feature extraction algorithm for vee-joints. As is evident from these figures, the algorithm yields incorrect results on the distorted range image $\{\mathbf{S}_i(k)\}$. However, on the transformed range image $\{\mathbf{P}_i(k)\}$, the feature points are correctly identified. The transformed range image $\{\mathbf{P}_i(k)\}$ describes the kth scan in the $(k-1)^{th}$ Seam coordinate frame. This frame is closer to the actual scene of the scan and so the transformed range image has less distortion, making it segmentable. To transform the range image from the Sensor coordinate frame to the Interpretation coordinate frame, the the kth range image is described in the world space using two alternative ways yielding the following identity,

$$[\mathbf{R}(k-n)]\left[\mathbf{T}_2\,\mathbf{S}\right]\left\{\mathbf{S}_i(k)\right\}=[\mathbf{M}(k)]\left\{\mathbf{P}'_i(k)\right\} \qquad (5.6)$$

91

The left side of the equation describes the world coordinates of the surface points on the seam as obtained through the range sensor attached to the robot arm while the right side describes the same points in the (k-1)th Seam coordinate frame. This implies,

$$\{P'_i(k)\} = [M(k)]^{-1}[R(k-n)][T_2 S]\{S_i(k)\} \qquad (5.7)$$

The initial Interpretation Coordinate frame $[M(0)]$ (for k = 0), is computed using a drive transformation relating the torch locations $[R(-n)]$ and $[R_I]$, which are defined by the sensor positioned at the seam start and the torch positioned at the seam start, respectively. It is given by

$$[M(0)] = [R(-n)][D(l)][R_2 C], \qquad (5.8)$$

where,

$$l = \frac{(n-1)}{n}$$ and n is the lag between the sensor and the torch.

Remark: Since the positioning of the sensor at the seam start is usually normal to the seam, the original range will normally be segmentable. However, for general startup conditions, the transformation of the range image from the image space to the Interpretation space increases the reliability. ♣

5.3.3 Bad Data Suppression and Data Filtering

The range sensor is susceptible to noise from a variety of sources, including, arc noise, specular reflection from the surface, and multiple reflections within the vee-joint. The adverse effects of the arc noise are mitigated through the use of a selective optical interference filter centered around the wavelength of the laser scanning beam (780 nanometers in our implementation). In this context, bad data suppression primarily concerns the removal of noisy data due to multiple reflections in the vee-joint and spurious surface reflections. Due to the geometry of a vee-jointed seam, any multiple reflections on the CCD array reduces the true range value. The heuristics for bad data suppression is based on this knowledge.

The *Smoothing Filter* alleviates the effects of high frequency noise in the range image (Morrison 1969). The noise source, which is primarily spurious surface reflections, is of high frequency and therefore, a low–pass filter is used. This filter uses weighted averages of the data on a moving window of size $(2m+1)$. Operating on any kth range image, the transformed and smoothed range value at any ith beam position is given by

$$z'_i(k) = \xi_{-m} \cdot z_{(i-m)}(k) + \ldots + \xi_0 \cdot z_i(k) + \ldots + \xi_m \cdot z_{(i+m)}(k) \qquad (5.9)$$

$$for \quad i = 0 \ldots r-1$$

with

$$\sum_{-m}^{m} \xi_j = 1.0, \quad -m \le j \le m$$

where,

z_i: original range value of the ith point in the range image;

ξ_j: weight associated with the $(i+j)$th range value.

The larger the window size (defined by a large value of m) greater is the effect of neighboring points on the filtered value. The effective contribution of each point is controlled by the weighting factor ξ_j . The value of the window size $(2m+1)$ and the weighting factor ξ_j is selected based on the seam geometry. For experimental purposes,

$$m = 1,$$
$$\xi_{-1} = 0.125 \qquad \xi_0 = 0.75 \qquad \xi_{+1} = 0.125$$
$$z_{-1} = z_0 \qquad z_r = z_{(r-1)}$$

This gives more weight to the value of the current point as opposed to its neighbors.

5.3.4 Feature Extraction and Seam Recognition

The segmented range image of a vee-joint is characterized by three feature points: the two outer points describe the edges while the center point describes the root of the joint. During seam tracking, the location of the sensor relative to the seam changes from scan to scan causing the view-vector to move away from the normal to the seam. This results in many range images being nonsegmentable causing the feature extraction algorithm to incorrectly identify the feature points of the seam. Figures 5.3(a)–5.7(a) show the results of the feature extraction algorithm on the original range image in the Sensor coordinate frame. On the other hand, all range images after transformation into the Interpretation coordinate frame,

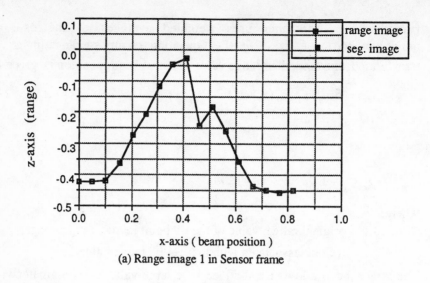

(a) Range image 1 in Sensor frame

Figure 5.3 (a) Original range image of a vee-groove in Sensor frame [S].
Feature extraction algorithm identifies four feature points.

bad data suppression, and noise filtering can be correctly segmented as seen
in Figures 5.3(b)–5.7(b).

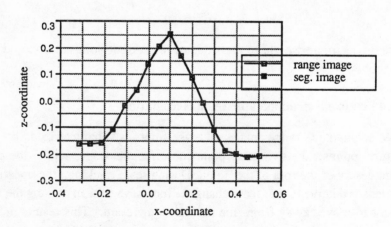

(b) Range image 1 in Interpretation frame

Figure 5.3(b) Transformed, filtered range image in Interpretation frame.
Feature extraction algorithm identifies three feature points.

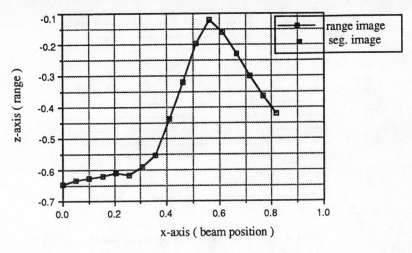

(a) Range image 2 in Sensor coordinate frame

Figure 5.4(a) Effect of excess view-vector angle on range image. Feature extraction algorithm identifies three incorrect feature points.

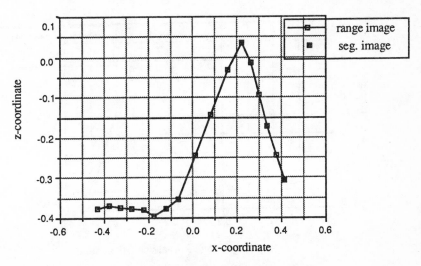

(b) Range image 2 in Interpretation frame

Figure 5.4(b) Above range image transformed in Interpretation frame. Feature extraction algorithm correctly identifies the seam's feature points.

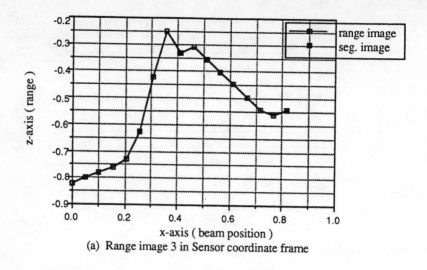

(a) Range image 3 in Sensor coordinate frame

Figure 5.5(a) Range image of a vee-groove affected by noise from specular reflection. Note the incorrect location of the root point, although all feature points are identified.

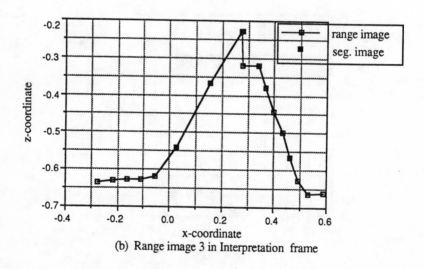

(b) Range image 3 in Interpretation frame

Figure 5.5(b) Above range image in the Interpretation frame. Note the new locations of the feature points identified by the feature extraction algorithm.

96

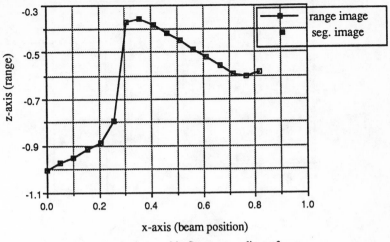

(a) Range image 4 in Sensor coordinate frame

Figure 5.6(a) Range of a vee-groove affected by excess view-vector angle.
Note the absence of the right edge point.

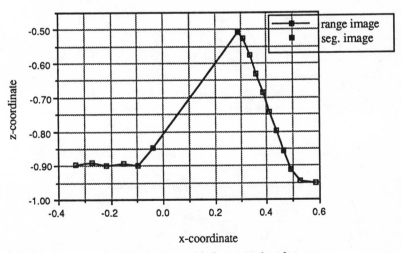

(b) Range image 4 in Interpretation frame

Figure 5.6(b) Above range image in the Interpretation frame. Note the
correct identification of all the feature points in this image.

97

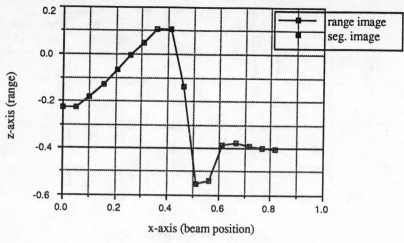

(a) Range image 5 in Sensor coordinate frame

Figure 5.7(a) Range image of a vee-groove in Sensor frame. Note the
effect of noise from specular reflection in locating the feature
points.

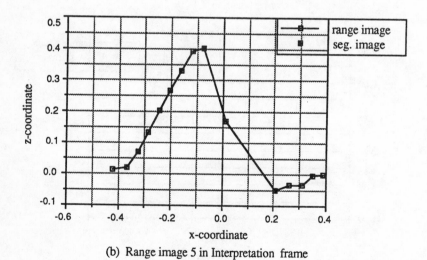

(b) Range image 5 in Interpretation frame

Figure 5.7(b) Above range image in the Interpretation frame. The
smoothing filter removes the effect of noise (due to specular
reflection) from the range image.

98

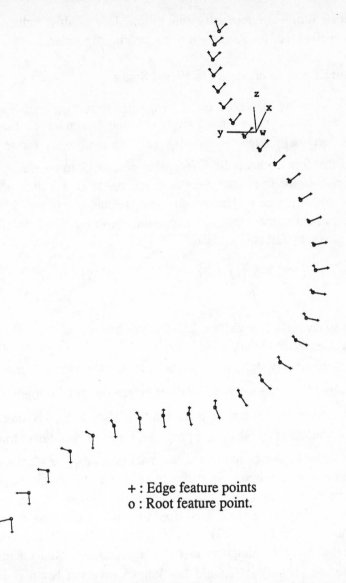

+ : Edge feature points
o : Root feature point.

Figure 5.8 Sequence of vee-groove feature points in world space.

Figure 5.8 shows a sequence of feature points extracted from the range images of a 3D seam in the Base coordinate frame. The original images were obtained by scanning a vee-groove milled into an aluminum plate, approximately one inch thick with a vee-groove approximately one-half inch deep. The points marked + (plus sign) represent the edgepoints while those marked with ∘ (circle) represent the root points. This result is indicative of the normal operation of the Range image processing algorithm.

5.3.5. Feature Point Coordinates in World Space

The feature points as identified in the range image remain unchanged in both, the transformed range image $\{P'_i(k)\}$ in the Interpretation frame as well as the original range image $\{Q_j(k)\}$ in the Sensor frame. For example, if the first feature point is located at position 3 in the transformed range image, then the third scan point will still be the first feature point in any other coordinate frame. Hence, once the feature points are identified in the transformed range image, their coordinates in the Base coordinate frame are given by Equation (5.10),

$$\left\{w_j(k)\right\} = \left[R_{i(j)}(k)\right]\left[T_2S\right]\left[Q_j(k)\right], \quad j = 1...3, \qquad i(j) = 0...r-1 \quad (5.10)$$

where,

j: index to the feature points, j = 1...3 for vee-grooves;

i(j):   position of the jth feature point in the scan, i(j) = 0 ... r-1;

$\left\{w_j(k)\right\}$: coordinates of the feature points in the Base coordinate frame;

$\left[R_{i(j)}(k)\right]$: describes the robot endeffector in the Base coordinate frame while scanning the jth feature point, which is also the $i(j)^{th}$ beam point;

$\left[Q_j(k)\right]$: coordinates of the jth feature point in the Sensor coordinate frame.

During seam tracking, the lookahead range sensor while scanning the seam surface, is also moving with the torchtip. If the scantime is a significant portion of the total cycletime, then the motion of the torch-sensor assembly should be accounted for in the world coordinates of the feature points (refer to Figure 5.8). This requires the correct location of the endeffector at each scanpoint in the range image. This means, in Equation (5.10) $\left[R_{i(j)}(k)\right]$ should be known at each beam position corresponding to each jth feature point identified in the kth range image.

During the kth scan, the torch is moving from the starting location $[R(k-n)]$ to the destination location $[R(k-n+1)]$. The starting location $[R(k-n)]$ is recorded by the robot controller while the destination location $[R(k-n+1)]$ is the control input. If these two coordinate frames are related

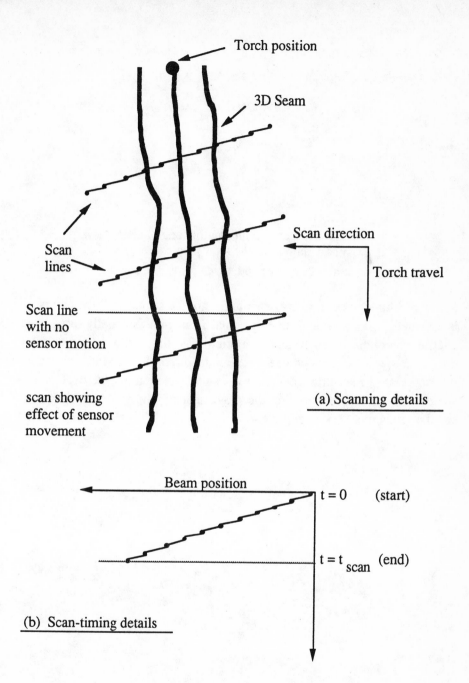

Torch position

3D Seam

Scan direction

Torch travel

Scan lines

Scan line with no sensor motion

(a) Scanning details

scan showing effect of sensor movement

Beam position

$t = 0$ (start)

$t = t_{scan}$ (end)

(b) Scan-timing details

Figure 5.9 Effect of sensor's motion on scanning during seam tracking.

by the drive transformation $[\mathbf{D}(l)]$, then the location of the endeffector $\left[\mathbf{R}_i\right]$ while scanning the ith point in the range image is given by

$$\left[\mathbf{R}_i\right] = [\mathbf{R}(k-n)][\mathbf{D}(l)], \qquad i = 0 \dots r-1, \tag{5.11}$$

where,

$$l = \left(\frac{i}{r}\right)\left(\frac{t_{scan}}{t_{cycle}}\right),$$

r is the total number of points in the range image,
t_{scan} is the time required for the total scan,
t_{cycle} is the control cycle period of seam tracking.

Remark: Although the time for scanning is not the same at all the points, it is assumed, for simplicity, that the total scan time is equally distributed over all scan points. Such an assumption has the advantage that only one time query at the end of the scan is required. Its validity is based on the fact that the scan time per point is a small fraction of the cycle time. Hence, the error in computing the movement of the sensor during such small periods should be negligible.   ♣

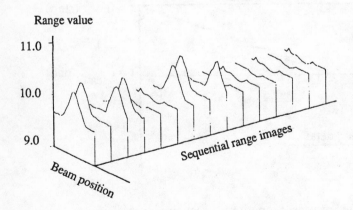

Figure 5.10 Sequence of Range Images from a vee-groove joint with tack welds

102

5.4 SEAM ENVIRONMENT MODEL & HIGH-LEVEL CONTROL

The range image interpretation stage outputs a model of the seam environment consisting of the history of the seam types encountered along the seam. Figure 5.10 shows a sample sequence of the range images input to the high-level controller while Table 5.1 describes the operation of the top-down feature interpretation algorithm in terms of the seam type identified, the number of iterations required for identification, and the rules applied in the process (refer to Section 3 in Chapter 3 for explanation of the rules). The model of the seam environment is used by the high-level controller to provide overall control of the welding process through adjustment of the weld parameters such as the tracking speed, welding voltage, etc.

Table 5.1 Performance of the heuristic approach to adaptively modeling the unstructured seam environment shown in Figure 5.10.

Step k	Image No	Initial Decision about seam type	Decision Modified ?	Rules Used
1	1	Vee-joint	No	Rules: 03, 11, 21
2	2	Vee-joint	No	Rules: 03, 11, 21
3	3	Vee-joint	No	Rules: 03, 11, 21
4	4	Presently Unknown	Tack (at k = 6)	Rules: 01, 02, 13
5	5	Presently Unknown	Tack (at k = 6)	Rules: 01, 02, 11
6	6	Vee-joint	No	Rules: 03, 12, 21
7	7	Vee-joint	No	Rules: 03, 11, 21
8	8	Vee-joint	No	Rules: 03, 11, 21
9	9	Presently Unknown	End-of-seam	Rules: 01, 02, 13
10	10	Presently Unknown	End-of-seam	Rules: 01, 02, 11
11	11	Presently Unknown	End-of-seam	Rules: 01, 02, 11
12	12	End-of-seam	--	Rules: 01, 02, 11, 14

The overview of high-level control involving seam environment modeling and subsequent welding process control is shown in Figure 5.11. Although the discussion on welding process control is outside the scope of this monograph, a few possible torch-seam interaction scenarios are mentioned for the sake of completeness:

1. By online monitoring of the seam's radius of curvature in the scanning and the welding region, the tracking speed can be adjusted in realtime. While scanning rapidly changing seams with small radius of curvature, a reduced tracking speed is preferred in order to have closely spaced scans for a detailed seam model. The speed should also be reduced while tracking seams with small radius of curvature, in order to prevent high accelerations of the robot arm joints.

2. The high-level controller can interact with an online welding knowledge base (KB) to perform realtime process control. For instance, changes in the tracking speed require corresponding changes in the wirefeed rate to properly fill the joint. The proper wirefeed rate can be known by consulting the welding KB. Similar decisions are

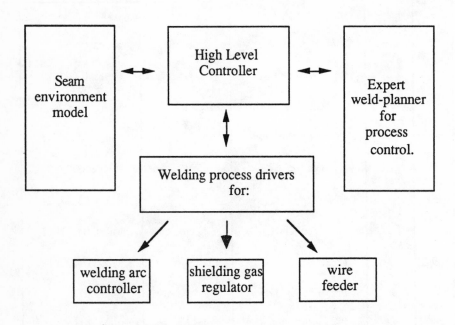

Figure 5.11 Architecture of the high-level controller for interacting with the seam environment during intelligent seam tracking.

104

required when the joint volume changes with changing seam geometry.

3. For tack welds along the seam, the high-level controller's response may be to either increase the welding current and burn through the tack weld, or to reduce the wirefeed rate when passing over the tack.

Remark: In the absence of an expert weld planner, some developers store torch-seam interaction details *a priori* by using a preteach stage (2nd generation system). The seam tracking system applies correction to the pretaught weld path in realtime and selects the correct weld process control parameters. This approach also has the advantage of knowing the robot arm joint accelerations *a priori* for controlling the welding speed without overloading the joint motors. The advantages and disadvantages of preprogrammed path systems have been discussed in Chapter 1. ♣

5.5 GEOMETRIC MODEL OF THE SEAM

The seam geometry is represented by a hybrid model consisting of both, a continuous model and a discrete model. The continuous model of the seam is a sequence of least squares cubic polynomials in parametric form representing the root curve. On the other hand, the discrete model of the seam is a sequence of Seam coordinate frames, $[C(k)]$ at the scanned points along the seam. Both models are generated from a sequence of N feature point sets $\{w_j(k), k = 1...N\}$, described in the Base coordinate frame. The two models complement each other in computing the control input to the robot for controlling (1) the torch's location and orientation and (2) the sensor's rotation about the torch axis.

5.5.1 Continuous Model of the Seam

The continuous model of the seam is required for two purposes. First, coordinates of root points can be interpolated anywhere along the seam from this model. This feature is particularly relevant since the range images of the seam may not be equidistant, while the torch positions during seam tracking have to be computed at equidistant locations. Second, the coordinates of the next scan location can be predicted from the continuous model. This information is necessary for orienting the sensor relative to the seam.

In the continuous model of the seam, only the root curve is represented by a sequence of three least squares cubic polynomials in parametric form. These cubic polynomials represent the x, y, and z coordinates of the root curve as functions of the independent variable s,which is the distance measured along the seam. Although, the physical seam can be represented by separate curves along the root and the other two edges, the seam's hybrid model adequately represents the seam for seam tracking purposes. Therefore, there is no need to represent the other two curves along the edges of the seam.

The choice of the least squares approach to modeling the seam function is based on the fact that any path-point generated is subjected to errors. The sources of errors include such factors as the resolution of the range image, the inherent inaccuracy in positioning/locating the robot's endeffector, etc. It is possible to construct either a polynomial or a cubic spline which fits the data exactly at each of the path-points. However, the disadvantage of such an approximation is that not only would it include all of the noise in the data but it would necessarily be a polynomial of a very high degree and would oscillate wildly, perhaps straying far from the immediate region which contains the data. Hence a better functional approximation would be one which tends to *smooth* the data and the *least squares polynomial approximation* is one such technique (Weeg and Reed 1966; Frober 1969).

The entire root curve along the seam is represented by a sequence of polynomials in parametric form, each defined over a small moving window, rather than a single polynomial. Over a small window, the seam is not expected to change rapidly, and hence a cubic polynomial adequately represents the root curve coordinates. For the kth set of cubic polynomials in the sequence (corresponding to the kth range image), the twelve coefficients $\left\{a_i(k), b_i(k), c_i(k) \quad i = 0, 1, 2, 3\right\}$ are computed recursively using the least squared error method (refer Appendix B at end of chapter). The computations are performed on a set of feature points $\left\{ \mathbf{w}_2(k-m), \dots \mathbf{w}_2(k), \dots \mathbf{w}_2(k+m) \right\}$ describing the root of the vee-jointed seam over a moving window of size $(2m+1)$. The kth approximating polynomial in parametric form is given by

106

$$G_{x,k}(s) = a_0(k) + a_1(k)(s - d(k)) + a_2(k)(s - d(k))^2 + a_3(k)(s - d(k))^3$$

$$G_{y,k}(s) = b_0(k) + b_1(k)(s - d(k)) + b_2(k)(s - d(k))^2 + b_3(k)(s - d(k))^3$$

$$G_{z,k}(s) = c_0(k) + c_1(k)(s - d(k)) + c_2(k)(s - d(k))^2 + c_3(k)(s - d(k))^3$$

$$d(k) = d(k-1) + \| w_2(k) - w_2(k-1) \|, \quad -m \le k \le \infty$$

$$(5.12)$$

where,

$\quad s \quad$ is the distance along the seam;

$\quad d(k) \quad$ is the distance of the kth root point along the seam.

To compute the coordinates of the root point at a distance s along the seam, the (k^{th}) polynomial selected in the sequence is such that s preferably lies between $d(k)$ and $d(k+1)$. However, the same polynomial can equally well be used to compute the root point coordinates over the entire data window of size $(2m+1)$, i.e., $d(k-m) \le s \le d(k+m)$

Remark: For selecting the window-size $(2m+1)$, it should be noted that for a rapidly changing seam geometry, a smaller data window represents the seam more accurately than a larger window. Since a large data window describes a large section of the seam, a cubic polynomial may not be of sufficiently high degree to accurately represent the seam over the entire window. Consequently, a large window size may introduce a significant error in the approximation. On the other hand, the seam is not expected to change rapidly within a small data window, thus allowing the least squares cubic polynomial to adequately represent the corresponding seam function. Appendix B at the end of this chapter discusses in detail the least squares approximation method and the error in using different data window sizes (ranging from 5 to 13) to approximate the x, y, and z coordinates of the root curve. These experiments show that the error increases as the data window size is increased from 5 to 13 and therefore, the smaller data window of size 5 is selected because the resulting curves provide a better fit for the seam's root curve. ♣

Figure 5.12 shows the continuous model for the vee-groove, computed recursively from a moving data window of size 5. While a minimum of four points is required to describe a least squared error cubic polynomial, a choice of five points has the advantage of providing symmetry about the central (k^{th}) point in the window. The seam tracking parameters in this

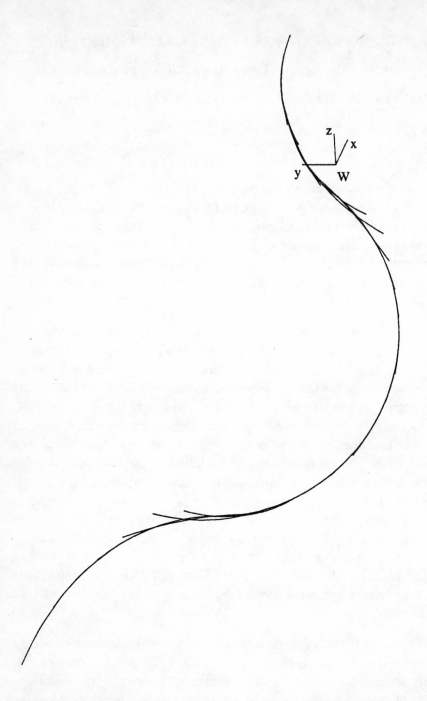

Figure 5.12 Sequence of x, y, and z cubic polynomials along the root curve representing the continuous model of the vee-grooved seam.

case are tracking speed = 60 in/min and control-cycle period = 1000 msecs. The tracking accuracy can be further improved by using a lower speed which allows the scans to be placed closer.

5.5.2 Discrete Model of the Seam

As mentioned earlier, the discrete model of the seam is represented as a sequence of Seam coordinate frames $\{[C(0)]...[C(k)]\}$ (refer to Figure 4.1 for illustration of various coordinate frames). The discrete model of the seam serves two purposes. First, the kth Seam coordinate frame serves as the Interpretation coordinate frame in the (k+1)th cycle. The role of the Interpretation Coordinate Frame in the analysis of range images and in identifying the feature points has been discussed in the previous sections. Second, while the position of the torch is computed from the continuous model, the orientation of the torch at any point along the seam is computed from the discrete model.

5.5.2.1 Computing the kth Seam Coordinate Frame. To compute the seam coordinate frame $[C(k)]$ from the world coordinates of the three vee-joint feature points, the following procedure is used.

Let $\{w_1(k), \quad w_2(k), \quad w_3(k)\}$ be the world coordinates of the three feature points extracted from the kth range image, where $w_1(k)$ and $w_3(k)$ are the joint edge points, $w_2(k)$ is the root point, and

$$w_j(k) = \left[w_{jx}(k) \quad w_{jy}(k) \quad w_{jz}(k) \quad 1.0\right]^T, \qquad j = 1, 2, 3 \qquad (5.14)$$

Let the kth Seam coordinate frame $[C(k)]$ be given by the (4 x 4) homogenous transformation matrix (Denavit and Hartenburg 1955; Paul 1977; Pieper 1968).

$$[C(k)] = \begin{bmatrix} n_x(k) & o_x(k) & a_x(k) & p_x(k) \\ n_y(k) & o_y(k) & a_y(k) & p_y(k) \\ n_z(k) & o_z(k) & a_z(k) & p_z(k) \\ 0 & 0 & 0 & 1 \end{bmatrix} \qquad (5.15)$$

where

$n(k)$ is unit vector along the x direction of the kth Seam coordinate frame in the Base frame,

109

$\mathbf{o}(k)$ is unit vector along the y direction of the kth Seam coordinate frame in Base frame,

$\mathbf{a}(k)$ is unit vector along the z direction of the kth Seam coordinate frame in the Base frame,

$\mathbf{p}(k)$ is Seam coordinate frame's origin in the Base frame.

These four component vectors of the seam coordinate frame are computed from the three feature points as follows:

Origin $\mathbf{p}(k)$: The kth Seam coordinate frame's origin is described by the coordinates of the feature point representing the root in the kth segmented range image,

$$\mathbf{p}(k) = \begin{bmatrix} p_x(k) & p_y(k) & p_z(k) & 1.0 \end{bmatrix}^T \qquad (5.16)$$
$$= \begin{bmatrix} w_{2x}(k) & w_{2y}(k) & w_{2z}(k) & 1.0 \end{bmatrix}$$

Orientation Vector $\mathbf{o}(k)$: The unit vector $\mathbf{o}(k)$ represents the seam direction and is described by the slope vector along the root curve in the continuous model, at $s = d(k)$. For the kth range image, the last completed root curve segment is (k-m). Hence, mathematically

$$\mathbf{o}(k) = \frac{\left. \dfrac{\partial G_{x,k-m}}{\partial s} \right|_{s=d(k)} \mathbf{i} + \left. \dfrac{\partial G_{y,k-m}}{\partial s} \right|_{s=d(k)} \mathbf{j} + \left. \dfrac{\partial G_{z,k-m}}{\partial s} \right|_{s=d(k)} \mathbf{k}}{\left\| \left. \dfrac{\partial G_{x,k-m}}{\partial s} \right|_{s=d(k)} \mathbf{i} + \left. \dfrac{\partial G_{y,k-m}}{\partial s} \right|_{s=d(k)} \mathbf{j} + \left. \dfrac{\partial G_{z,k-m}}{\partial s} \right|_{s=d(k)} \mathbf{k} \right\|} \qquad (5.17)$$

where,

$G_{*,*}$ is the polynomial describing the root curve as defined in Equation 5.12;

$\mathbf{i}$ is a unit vector along the x-axis of the base coordinate frame;

$\mathbf{j}$ is a unit vector along the y-axis of the base coordinate frame;

$\mathbf{k}$ is a unit vector along the z-axis of the base coordinate frame.

The distance $d(k)$ along the seam that is associated with the kth range image is recursively obtained as

$$d(k) = d(k-1) + \| \mathbf{w}_2(k) - \mathbf{w}_2(k-1) \| \qquad (5.18)$$

Approach Vector $\mathbf{a}(k)$: The unit vector $\mathbf{a}(k)$ represents the direction normal to the seam. It is computed from the bisection vector $\mathbf{b}(k)$ of

110

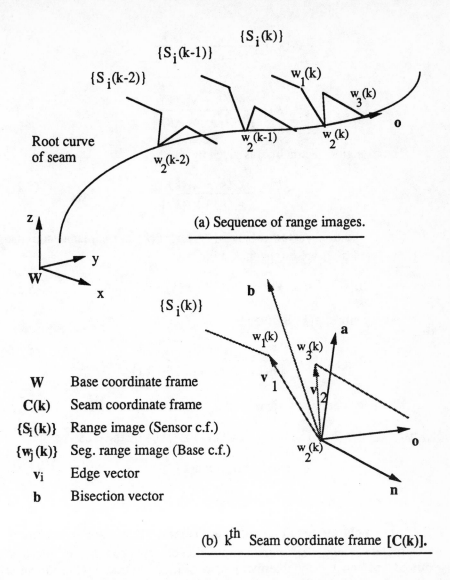

(a) Sequence of range images.

W	Base coordinate frame
C(k)	Seam coordinate frame
{S_i(k)}	Range image (Sensor c.f.)
{w_j(k)}	Seg. range image (Base c.f.)
v_i	Edge vector
b	Bisection vector

(b) k^{th} Seam coordinate frame **[C(k)].**

Figure 5.13 Computation of the Discrete Model of the Seam.

the vee-joint as shown in Figure 5.13. Let vector $b(k)$ bisect the two edge vectors $v_1(k)$ and $v_3(k)$, where

$$v_1(k) = \frac{w_1(k) - w_2(k)}{\| w_1(k) - w_2(k) \|}, \tag{5.19}$$

$$v_3(k) = \frac{w_3(k) - w_2(k)}{\| w_3(k) - w_2(k) \|} \tag{5.20}$$

The bisection vector $b(k)$ is given by

$$b(k) = \frac{v_1(k) + v_3(k)}{\| v_1(k) + v_3(k) \|} \tag{5.21}$$

Since the unit vectors $o(k)$, $b(k)$, and $a(k)$ are coplanar and also $o(k)$ and $a(k)$ are orthogonal, it implies

$$b(k) = (\, b(k) \bullet a(k) \,)a(k) + (\, b(k) \bullet o(k) \,)o(k) \tag{5.22}$$

and therefore, $a(k)$ is given by

$$a(k) = \frac{b(k) - (b(k) \bullet o(k))o(k)}{\| b(k) - (b(k) \bullet o(k))o(k) \|} \tag{5.23}$$

where the operator $\bullet$ signifies dot product of two vectors.

<u>Normal Vector</u> $n(k)$: The unit vector $n(k)$ is given by the cross-product

$$n(k) = o(k) \times a(k) \tag{5.24}$$

to form a right-handed coordinate system.

5.5.2.2 Interpolating Seam Coordinate Frames. Just as the root points can be interpolated along the seam, interpolation of seam coordinate frames is required as well. Any intermediate coordinate frame $[C(i)]$ located between $[C(k)]$ and $[C(k+1)]$ can be interpolated through a drive transformation $[D(l)]$ relating the two coordinate frames. $[C(i)]$ is based on the relative motion parameter (l), which is given by

$$l = \frac{\| d_i - d(k) \|}{\| d(k+1) - d(k) \|}, \qquad 0 \le l \le 1, \tag{5.13}$$

where,

112

d_i is the distance of the $[C(i)]$ coordinate frame's origin along the seam.

During the ith step in seam tracking, the torch is positioned at a distance d_i from the start of the seam. Note that the torch orientation in the ith step and the orientation of the intermediate coordinate frame $[C(i)]$ are related through the fixed coordinate transformation $[C_2R]$.

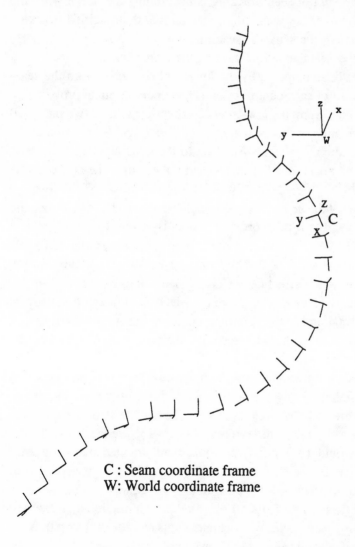

C : Seam coordinate frame
W: World coordinate frame

Figure 5.14 Sequence of seam coordinate frames representing the discrete model of the seam.

113

Figure 5.14 shows the discrete model of the seam consisting of a sequence of Seam coordinate frames computed using the above procedure. The y-axis of the Seam coordinate frame is along the direction of the seam while the z-axis is normal to the vee-groove.

5.6 ROBOT MOTION CONTROL (WELD PATH GENERATION)

The basic requirement of seam tracking is positioning the torch with the correct orientation over the seam while maintaining the specified tracking speed. Additionally, the sensor's scanning window should be properly positioned over the seam to allow tracking the seam ahead.

The torch path is computed by the low-level controller from the seam models. At the start of each control cycle (k), the control input provided to the robot controller from the low-level controller consists of two parts: (i) position and orientation of the torch at step (5 DOF), and (ii) rotation of the sensor about the torch axis (1 DOF). Due to the dynamics of the system, this destination is not immediately reached but requires the entire cycle time (t_{cycle}). Hence, the robot controller input at any k^{th} instant is defined in terms of the desired destination location at the $(k+1)^{th}$ instant. The position of the Torch-tip coordinate frame is given by the x, y, and z coordinates (3 DOF). The orientation of the torch is described in terms of its rotations about the x and the y axes (2 DOF). Since the rotation about the z-axis does not affect the welding process, it is used to control the rotation of the sensor about the torch axis (1 DOF). Although these inputs are computed separately, their implementation has to be carried out simultaneously because both, the torch and the sensor are mounted on the same robot endeffector.

An easy way to understand the compounding of these two separate input transformations is by initially considering only the input for positioning and orienting the torch (5 DOF) at the $(k+1)^{th}$ step, assuming no rotation of the sensor about the torch axis (1 DOF). Hence at the $(k+1)^{th}$ step, the torch would be properly positioned and oriented above the seam but the sensor may not be on the seam. Now consider only the rotation of the sensor about the torch axis for positioning the scanning window over the seam for the $(k+n+1)^{th}$. This is the desired pose of the robot at the step $(k+1)$. When completed, these two separate inputs can be combined through a compound transformation (Paul 1981) given by

$$[U(k)] = [Y(k+1)][Z(k+1)][C_2R] \qquad (5.25)$$

114

where,

$[U(k)]$: is a (4 x 4) transformation representing the compounded control input (6 DOF) to the robot controller given at the start of the kth cycle for a desired pose at the (k+1)th step.

$[Y(k+1)]$: is a (4 x 4) transformation matrix, relative to the Base coordinate frame, representing the desired position and orientation of the torch tip (5 DOF), at the $(k+1)^{th}$ destination location. Note that $[Y(k+1)]$ is related to the seam and the transformation $[Y(k+1)][C_2R]$ is necessary to compute the proper position the torch over the seam.

$[Z(k+1)]$: is a (4 x 4) transformation matrix, relative to the $[Y(k+1)]$ coordinate frame, representing the desired rotation of the sensor about the torch's z-axis (1 DOF) at the (k+1)th destination location.

5.6.1 Position and Orientation of the Torch (5 DOF)

Two separate conditions during seam tracking are considered while computing the control input. First, the initial path when the sensor has just begun tracking the seam but the torch is still outside the seam start. Second, the normal seam tracking operation past the starting position. During the seam tracking operation, the sensor is leading the torch by n steps. This allows the buffering of the $[Y(k+1)]$ transformations along the path for compounding with $[Z(k+1)]$ when the torch is ready to move to that location.

5.6.1.1 Initial Path (for k < 0). The torchtip is moved along the initial path to reach the start of the seam. The desired position and orientation of the torch $[Y(k+1)]$ during the initial path is computed using a drive transformation $[D(l)]$ relating the two coordinate frames $[C(-n)]$ and $[C_I]$ such that

$$[Y(k+1)] = [C(-n)][D(l)] \qquad (5.26)$$

where,

$$l = \frac{(n+k)}{n}, \qquad 0 \le l \le 1, \qquad -n < k < 0,$$

n is the lag between the sensor and the torch.

115

The coordinate frame $\begin{bmatrix} C_I \end{bmatrix}$ is derived from the torch-tip location at the seam start $\begin{bmatrix} R_I \end{bmatrix}$ while $[C(-n)]$ is derived from the torch-tip location $[R(-n)]$ (when the sensor is at the seam start).

$$[C(-n)] = [R(-n)]\begin{bmatrix} R_2 C \end{bmatrix} \qquad (5.27)$$

$$[C_I] = [R_I]\begin{bmatrix} R_2 C \end{bmatrix}$$

The computation of the sensor's rotation about the torch axis during both, the initial-path tracking and normal seam tracking remains unchanged and is discussed in Section 5.6.2.

5.6.1.2 Seam Tracking (k ≥ 0). For a general seam tracking algorithm where the tracking speed $V_{weld}(i)$ can vary during any cycle, the torch travel $\bar{d}(k)$ from the seam start is given by

$$\bar{d}(k) = \sum_{i=1}^{k} d_{cycle}(i) \qquad (5.28a)$$

$$d_{cycle}(i) = V_{weld}(i) \cdot t_{cycle}$$

However, in our system this has been simplified by using a constant seam tracking speed. This requires that the torch be positioned at equidistant locations along the seam since the fixed tracking speed (V_{weld}) and constant cycle time (t_{cycle}) result in a constant torch travel per cycle (d_{cycle}). In k cycles, the torch tip should have traveled a distance $\bar{d}(k)$ from the start of the seam, where

$$\bar{d}(k) = k \cdot d_{cycle} \qquad (5.28b)$$

$$= k \cdot V_{weld} \cdot t_{cycle}$$

with the distance $\bar{d}(k)$ being measured along the seam.

The position and orientation of the torch at each step is computed as follows.

Position: The coordinates $\bar{p}(k+1)$ of the desired torch tip location at the (k+1)th instant, where

116

$$\bar{\mathbf{p}}(k+1) = \begin{bmatrix} \bar{p}_x(k+1) & \bar{p}_y(k+1) & \bar{p}_z(k+1) \end{bmatrix}^{\mathrm{T}} \tag{5.29}$$

are computed as a function of $\bar{d}(k+1)$ from the continuous model of the seam. For $d(i) \le \bar{d}(k+1) \le d(i+1)$

$$\bar{p}_x(k+1) = G_{x,i}(s)\big|_{s=\bar{d}(k+1)}$$
$$= a_0(i) + a_1(i)\tilde{d}(k+1) + a_2(i)\tilde{d}^2(k+1) + a_3(i)\tilde{d}^3(k+1)$$
$$\bar{p}_y(k+1) = G_{y,i}(s)\big|_{s=\bar{d}(k+1)} \tag{5.30}$$
$$= b_0(i) + b_1(i)\tilde{d}(k+1) + b_2(i)\tilde{d}^2(k+1) + b_3(i)\tilde{d}^3(k+1)$$
$$\bar{p}_z(k+1) = G_{z,i}(s)\big|_{s=\bar{d}(k+1)}$$
$$= c_0(i) + c_1(i)\tilde{d}(k+1) + c_2(i)\tilde{d}^2(k+1) + c_3(i)\tilde{d}^3(k+1)$$

where,

$$\tilde{d}(k+1) = \bar{d}(k+1) - d(i) \tag{5.31}$$

and $G_{*,*}$ is defined in Equation 5.12.

Orientation: The orientation of the torch tip is computed using the discrete model of the seam. If $d(i) \le \bar{d}(k+1) \le d(i+1)$ then the orientation can be derived using a drive transformation $[D(l)]$ relating the known seam coordinate frames $[C(i)]$ and $[C(i+1)]$. The coordinate transformation $[\hat{\mathbf{Y}}(k+1)]$ representing the desired orientation of the torch at the $(k+1)$th instant is then given by

$$[\hat{\mathbf{Y}}(k+1)] = [C(i)][D(l)], \qquad (k+1) \ge 0 \tag{5.32}$$

where,

$$l = \frac{\bar{d}(k+1) - d(i)}{d(i+1) - d(i)}, \qquad 0 \le l \le 1 \tag{5.33}$$

Remark: The orientation of the torch-tip is computed by linearly interpolating a coordinate frame between two adjacent seam coordinate frames, using the drive transformation $[D(l)]$. Although the orientation is

117

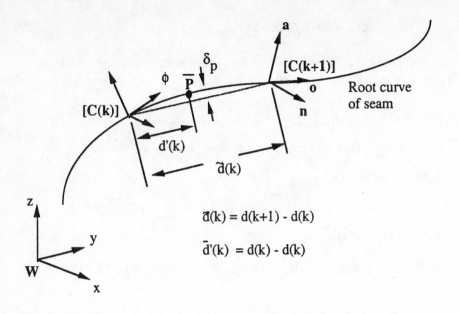

Figure 5.15 Simplified model of the seam for approximating the torch
orientation at location $\overline{\mathbf{p}}(k)$ at step (k) during seam tracking.

based on a linear interpolation between points [k, k+1], the torch position
is computed from the continuous model which is non-linear (refer to
Figure 5.15). Hence, the orientation and the position are computed at two
different locations along the seam. For these two locations to be close, the
error δ_p should be small implying that the seam's radius of curvature should
be greater than a certain threshold for a specified tracking speed. Further
details about specifying the minimum radius of curvature of the seam are
discussed in Chapter 7.   ♣

In the above (4 x 4) coordinate transformation $\left[\, \hat{\mathbf{Y}}(k+1)\,\right]$, where

$$\left[\hat{\mathbf{Y}}(k+1)\right] = \begin{bmatrix} n_x(k+1) & o_x(k+1) & a_x(k+1) & w_x(k+1) \\ n_y(k+1) & o_y(k+1) & a_y(k+1) & w_y(k+1) \\ n_z(k+1) & o_y(k+1) & a_y(k+1) & w_z(k+1) \\ 0 & 0 & 0 & 1 \end{bmatrix} \quad (5.34)$$

the first three columns represent the desired orientation of the torch. In this
matrix, the fourth column should be replaced by the desired position of the

118

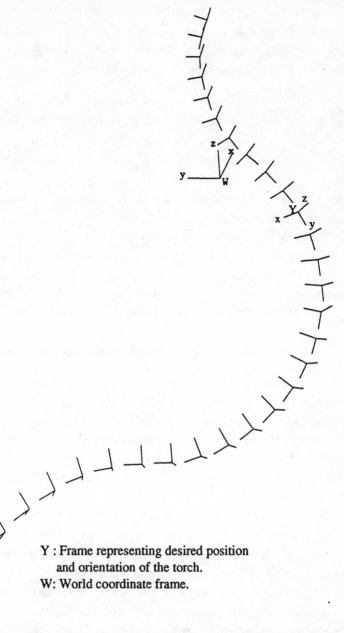

Y : Frame representing desired position
and orientation of the torch.
W: World coordinate frame.

Figure 5.16 Sequence of $\left[\,\mathbf{Y}(k+1)\,\right]$ coordinate frames describing the
desired position and orientation (5 DOF) of the torch along the
seam. Note the z-axis of this frame is normal to the vee-groove.

119

torch in order to obtain the coordinate transformation $[\mathbf{Y}(k+1)]$. This coordinate frame then represents both, the desired position and the desired orientation of the torch at the (k+1)[th] instant. By replacing the fourth column of the matrix in Equation (5.34) by $\overline{\mathbf{p}}(k+1)$, we get

$$[\mathbf{Y}(k+1)] = \begin{bmatrix} n_x(k+1) & o_x(k+1) & a_x(k+1) & \overline{p}_x(k+1) \\ n_y(k+1) & o_y(k+1) & a_y(k+1) & \overline{p}_y(k+1) \\ n_z(k+1) & o_z(k+1) & a_z(k+1) & \overline{p}_z(k+1) \\ 0 & 0 & 0 & 1 \end{bmatrix} \quad (5.35)$$

where the desired position $\overline{\mathbf{p}}(k+1)$ is computed from the continuous model of the seam and given by Equation 5.29.

Figure 5.16 shows a sequence of $[\mathbf{Y}(k+1)]$ coordinate frames in world space as computed using the above procedure. These coordinate frames are equidistant along the seam being tracked. The parameters used in the seam tracking process are: tracking speed V_{weld} = 60 inches/min, and control-cycle period t_{cycle} = 1000 milliseconds.

5.6.2 Computing the Sensor's Rotation about the Torch Axis (1 DOF)

A controlled rotation of the sensor about the torch axis is necessary to position the scanning window over the seam. Thus, while the control input supplied to the robot controller at the start of the kth cycle positions and orients the torch over the seam, as desired at the (k+1)[th] step, it should also orient the sensor over the seam to acquire the (k+n+1)[th] range image. To have the input ready at the start of kth cycle, the control input is computed during the interval [k-1, k] from the available information about the seam, specifically, root point coordinates $\{\mathbf{w}_2(0) \dots \mathbf{w}_2(k+n-1)\}$.

During the kth step, the last range image scanned is $k_f = k+n-1$. To compute the rotation angle, it is necessary to predict the location of the (k+n+1)[th] scan, which is a two-step-ahead prediction. The prediction is based on the continuous model of the seam, specifically, the last set of polynomials in the sequence, i.e. the $(k_f\text{-}m)$[th] polynomials. These polynomial coefficients are computed from the data window $\{\mathbf{w}_2(k_f - 2m) \dots \mathbf{w}_2(k_f - m) \dots \mathbf{w}_2(k_f)\}$ of size (2m+1).

Let, $i_f = k_f - m$ be the last polynomial in the sequence and $\hat{d}(k_f + 2)$ be the distance of the predicted root point along the seam, where

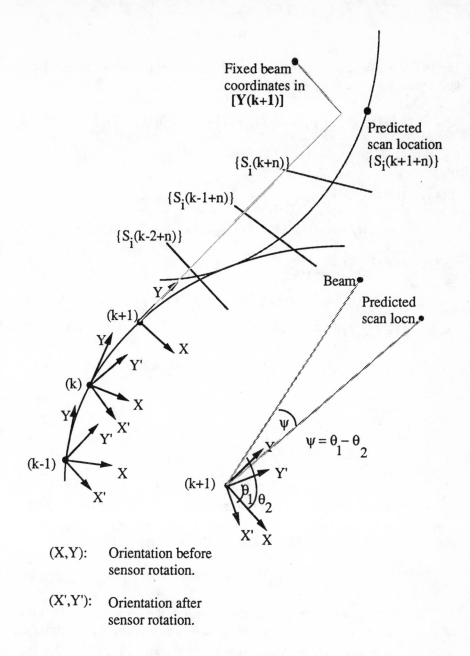

Figure 5.17 Computing the Sensor's Rotation about the Torch.

$$\hat{d}(k_f+2)=d(k_f)+2d_{cycle} \qquad (5.36)$$

Let the coordinates of the predicted root point relative to the Base coordinate frame be $\hat{p}(k_f+2)$, where

$$\hat{p}(k_f+2)=\begin{bmatrix} \hat{p}_x(k_f+2) & \hat{p}_y(k_f+2) & \hat{p}_z(k_f+2) \end{bmatrix}^T \qquad (5.37)$$

which is given by

$$\hat{p}_x(k_f+2)=G_{x,i_f}(s)\Big|_{s=\hat{d}(k_f+2)}$$
$$=a_0(i_f)+a_1(i_f)\tilde{d}(k_f+2)+a_2(i_f)\tilde{d}^2(k_f+2)+a_3(i_f)\tilde{d}^3(k_f+2)$$

$$\hat{p}_y(k_f+2)=G_{y,i_f}(s)\Big|_{s=\hat{d}(k_f+2)}$$
$$=b_0(i_f)+b_1(i_f)\tilde{d}(k_f+2)+b_2(i_f)\tilde{d}^2(k_f+2)+b_3(i_f)\tilde{d}^3(k_f+2)$$

$$\hat{p}_z(k_f+2)=G_{z,i_f}(s)\Big|_{s=\hat{d}(k_f+2)}$$
$$=c_0(i_f)+c_1(i_f)\tilde{d}(k_f+2)+c_2(i_f)\tilde{d}^2(k_f+2)+c_3(i_f)\tilde{d}^3(k_f+2)$$

$$(5.38)$$

where

$$\tilde{d}(k_f+2)=\hat{d}(k_f+2)-d(i_f) \qquad (5.39)$$

The sensor rotation (1 DOF) is computed about the z-axis of $[Y(k+1)]$ which represents the desired position and orientation of the torch at the $(k+1)$th instant. The coordinates of the two-step-ahead predicted root point $\hat{r}(k_f+2)$ relative to the $[Y(k+1)]$ coordinate frame, are given by

$$\hat{r}(k_f+2)=[Y(k+1)]^{-1}\hat{p}(k_f+2) \qquad (5.40)$$

To compute the angle of rotation, consider the projections of (i) the scanning laser beam, and (ii) the predicted root point $\hat{r}(k_f+2)$ in the xy-plane of the $[Y(k+1)]$ coordinate frame, as shown in Figure 5.17. The projection of the predicted root point $\hat{r}(k_f+2)$ in the xy-plane of $[Y(k+1)]$ coordinate frame is given by $\begin{bmatrix} \hat{r}_x(k_f+2), & \hat{r}_y(k_f+2) \end{bmatrix}^T$ and its angle with the x-axis is given as

122

$$\theta_1 = \tan^{-1}\left(\frac{\hat{r}_y(k_f+2)}{\hat{r}_x(k_f+2)}\right), \qquad -\pi \le \theta_1 \le +\pi \qquad (5.41)$$

On the other hand, the coordinates of the scanning beam's projection in the xy-plane of $[Y(k+1)]$ are fixed because the relationship between the Sensor coordinate frame and the Torch-tip coordinate frame is fixed. This relationship is given by the (4 x 4) coordinate transformation $[T_2S]$, the details of which were discussed in Chapter 4. Let the position of the beam in the scan center in the $[Y(k+1)]$ coordinate frame be $\mathbf{bm} = \begin{bmatrix} bm_x, & bm_y, & bm_z \end{bmatrix}^T$. The constant angle θ_2 that the beam's projection in the xy-plane of $[Y(k+1)]$ makes with the x-axis is given by

$$\theta_2 = \tan^{-1}\left(\frac{bm_y}{bm_x}\right), \qquad -\pi \le \theta_2 \le +\pi \qquad (5.42)$$

For a look-ahead range sensor, the beam and the predicted root point are expected to lie in either the first or the second quadrant of the xy-plane in $[Y(k+1)]$. This ensures that $0 \le \theta_1, \theta_2 \le +\pi$ and therefore the angle of rotation $\psi(k)$, required to position the scanning beam in line with the predicted scan location, is given by

$$\psi(k) = \theta_1 - \theta_2 \qquad (5.43)$$

The (4 x 4) transformation representing the rotation about the z-axis of the $[Y(k+1)]$ coordinate frame for correctly positioning the sensor over the seam is given by

$$[Z(k+1)] = \text{Rotation}(z, \psi(k)) \qquad (5.44)$$

$$= \begin{bmatrix} \cos(\psi(k)) & -\sin(\psi(k)) & 0 & 0 \\ \sin(\psi(k)) & \cos(\psi(k)) & 0 & 0 \\ 0 & 0 & 1 & 0 \\ 0 & 0 & 0 & 1 \end{bmatrix}$$

5.6.3 Pathpoint Coordinates (Control Input) to the Robot Controller

The control input $[U(k)]$ is given to the robot controller at the start of the kth cycle to (i) position the torch with the correct orientation, and

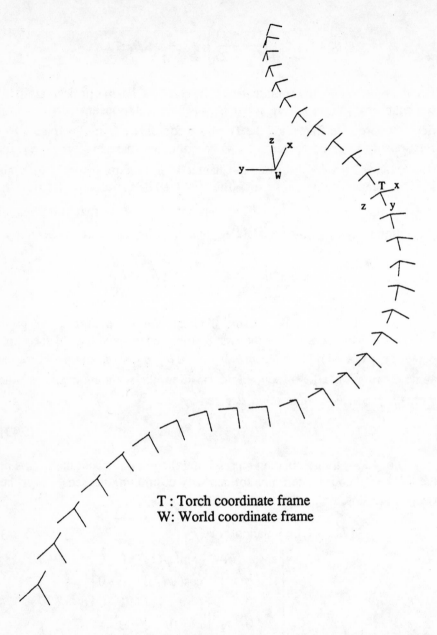

T : Torch coordinate frame
W: World coordinate frame

Figure 5.18 Sequence of $[U(k)]$ coordinate frames representing the control
input to the robot controller while tracking a vee-groove.

(ii) position the sensor over the seam, as desired at the $(k+1)^{th}$ instant. This input is described by a compound transformation given by

$$[\mathbf{U}(k)] = [\mathbf{Y}(k+1)][\mathbf{Z}(k+1)][\mathbf{C_2R}] \qquad (5.45)$$

Figure 5.18 shows the final transformation $[\mathbf{U}(k)]$ representing the torch's position and orientation and the sensor's rotation about the torch-axis, as computed using the above procedure. The fixed position of the beam in the $[\mathbf{Y}(k+1)]$ coordinate frame is $x=0.45$ inches, $y=6.917$ inches.

5.7 SUMMARY

Algorithms for seam tracking in an unstructured environment should primarily address issues of range image processing, high level overall control, and robot motion control. The basic range image processing techniques discussed in Chapter 3, have been enhanced for robustness to track seams in an unstructured environment. For example, the position of the sensor in relation to the seam is expected to change from scan to scan. In such a situation, the robustness of top-down feature extraction algorithm can be increased by transforming the range images from the sensor frame into the interpretation frame, thus ensuring that the transformed range images will be segmentable. Another enhancement necessary while computing the coordinates of the feature points in world space is to incorporate the motion of the torch-sensor assembly during scanning.

The robot motion control input during seam tracking is based on a hybrid model of the seam, which comprises of a continuous model and a discrete model. The continuous model consists of a sequence of cubic polynomials describing the x, y, and z coordinates of the root curve. The discrete model on the other hand, is a sequence of coordinate frames along the seam, describing the direction and orientation of the seam. The information input to the robot controller consists of (i) the position and orientation of the torch (5 DOF) and (ii) the rotation of the sensor about the torch (1 DOF). The torch position along the seam is computed from the seam's continuous model while the torch orientation is computed using the discrete model of the seam. On the other hand, the rotation of the sensor about the torch axis uses the computed position and orientation of the torch, and the predicted position of the root from the continuous model. These two inputs, when compounded together in the robot controller, control the torch and the sensor simultaneously during seam tracking.

Having discussed the details of robot motion control for seam tracking, the discussion now proceeds to an equally important aspect related to implementing seam tracking systems using off-the-shelf components. This is the subject of the next chapter.

APPENDIX B

B.1 LEAST SQUARES APPROXIMATION POLYNOMIALS

Approximation theory has provided many solution techniques to address the problem of fitting functions to given data and finding the "best" function in a certain class that can be used to represent the data. Certain approximation techniques provide functions that agree exactly with the given data. However, in the event that the given data is erroneous, these approximating functions introduce oscillations that were not originally present. Representing the root curve of the seam in the continuous model faces a similar problem. The range data acquired from the range sensor and the joint encoder information from the robot controller contain errors from a variety of sources (refer to Chapter 7 for sources of errors). In such a case, it is advantageous to find the "best" polynomial that could represent the given data even though it might not agree precisely with the data at any point.

The least squares polynomial approach to this problem involves determining the best approximating polynomial, where the error is the sum of the squared differences between the values on the approximation polynomial and the given data.

The general problem of approximating a set of data $\{(x_i, y_i) \mid i = 0,1,...M\}$, with a polynomial $P_n(x) = \sum_{k=0}^{n} a_k x^k$ of degree $n < M$, using the least squares procedure involves minimizing the error (E)

$$E = \sum_{i=0}^{M} \left(y_i - \sum_{k=0}^{n} a_k x_i^k \right)^2 \qquad (B.1)$$

with respect to the polynomial coefficients $a_0, a_1, ... a_n$. The least squares approach puts substantially more weight on a point that is out of line with the rest of the data but will not allow that point to completely dominate the approximation. In order for the error E to be minimized, it is necessary that $\partial E / \partial a_j = 0$ for each $j = 0,1,...n$. Thus for each j,

$$0 = \frac{\partial E}{\partial a_j} = -2 \sum_{i=0}^{M} y_i x_i^j + 2 \sum_{k=0}^{n} a_k \sum_{i=0}^{M} x_i^{j+k} \qquad (B.2)$$

This gives n+1 equations in n+1 unknowns, a_j, called the normal equations,

127

$$\sum_{k=0}^{n} a_k \sum_{i=0}^{M} x_i^{j+k} = \sum_{i=0}^{M} y_i x_i^j, \quad j = 0,1,\dots n. \qquad (B.3)$$

It is helpful to write out the equations as follows:

$$a_0 \sum_{i=0}^{M} x_i^0 + a_1 \sum_{i=0}^{M} x_i^1 + a_2 \sum_{i=0}^{M} x_i^2 + \cdots + a_n \sum_{i=0}^{M} x_i^n = \sum_{i=0}^{M} y_i x_i^0,$$

$$a_0 \sum_{i=0}^{M} x_i^1 + a_1 \sum_{i=0}^{M} x_i^2 + a_2 \sum_{i=0}^{M} x_i^3 + \cdots + a_n \sum_{i=0}^{M} x_i^{n+1} = \sum_{i=0}^{M} y_i x_i^1,$$

$$\vdots \qquad\qquad\qquad\qquad\qquad\qquad\qquad\qquad\qquad\qquad (B.4)$$

$$a_0 \sum_{i=0}^{M} x_i^n + a_1 \sum_{i=0}^{M} x_i^{n+1} + a_2 \sum_{i=0}^{M} x_i^{n+2} + \cdots + a_n \sum_{i=0}^{M} x_i^{2n} = \sum_{i=0}^{M} y_i x_i^n.$$

The normal equations can be solved to get the coefficients of the approximation polynomial $P_n(x)$. It can be shown that the normal equations always have a unique solution provided that the data $\{x_i, y_i\}$ for i=0, 1, ... M, are distinct. For cubic polynomials, $n = 3$ and the minimum value of $M = 4$.

The selection of M is an important issue in the continuous model of the seam. The continuous model comprises of three least squares cubic polynomials $P_x(s)$, $P_y(s)$, and $P_z(s)$ for the x, y, and z components of the root curve. These polynomials are a function of the distance (s) of any point along the root curve from the seam start. Since the entire seam function is unknown, we approximate short segments of the root curve using the least squares cubic polynomial. The continuous model of the entire root curve is thus, a sequence of least squares cubic polynomials with each polynomial representing a certain section of the root curve.

By selecting a large M (say 13), a large segment of the root curve is being approximated by the cubic polynomial. In this case, it is quite likely that the cubic polynomial is not of sufficient degree to correctly represent the entire seam section thereby introducing a large error in the approximation (refer to Figure B.1(a)-(c)).

Figures B.1(a)–(c) show the error introduced in approximating the three components (x, y, z) of the vee-groove with least squares cubic polynomials using two different window sizes (M=5, and M=13). This figure illustrates that in the absence of full knowledge about the root curve geometry, it is safer to use a smaller window (M=5) instead of a larger window (M=13).

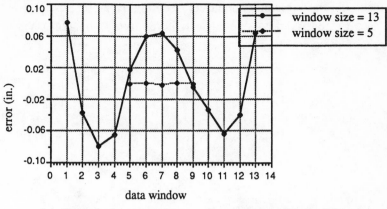

(a) Error in least sq. approx. for cubic x-polynomial

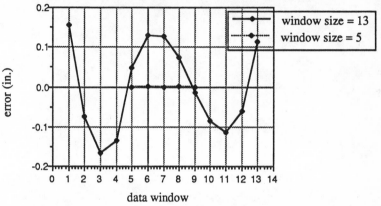

(b) Error in least sq. approx. for cubic y-polynomial

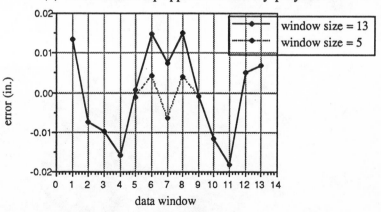

(c) Error in least sq. approx. for cubic z-polynomial

Figure B.1 (a)–(c) Selection criterion for the size of the moving data
window for the x, y, z cubic polynomials along the root curve.

Incidentally, the cubic polynomial representing the z-component is not much affected by the window size since it is relatively constant on the planar aluminum plate.

CHAPTER 6

IMPLEMENTATION CONCERNS IN SEAM TRACKING SYSTEMS

6.1 INTRODUCTION

The detailed analysis of the seam tracking process was discussed in Chapter 5. Equally important are the implementation concerns of this theory. To learn and weld the seam in the same pass, the operational requirements of this system include (i) interpreting the seam information from the weld joint image in realtime, (ii) adaptively generating models of both the seam geometry and the seam environment, and (iii) computing the control input to guide the welding torch and the sensor along the seam.

To provide this functionality, the seam tracker's architecture should feature hardware components and associated software modules to support a hierarchical control structure. The hardware in this system implementation includes a laser profiling gage (LPG), a six-axis, articulated-joint robot arm and its controller, a microcomputer to provide supervisory level control, and a part positioner.

A hierarchical control structure is the logical choice to implement the interaction of three distributed subsystems, i.e., the supervisory controller, the robot controller, and the range sensor. In addition, the seam tracking controller design should feature a two-level control scheme: a high-level controller and a low-level controller. The high-level controller adaptively models the unstructured seam environment and controls the torch-seam interaction. Hence, it should be able to make realtime decisions that affect the system's overall working. These functions are implemented using both rule-based heuristics and model-based reasoning techniques. Although, the welding aspects of the system are outside the scope of this discussion, it would suffice to mention that a welding knowledge base can be interfaced through the high-level controller.

The low-level controller, on the other hand, is concerned with robot motion control. Its function primarily includes computing the position and orientation of the torch and the sensor over the seam. As such, low-level controller tasks include, range image interpretation, adaptive modeling of the seam geometry, computing the control input from the seam model to position both torch and sensor over the seam, and communicating these inputs to the robot controller during seam tracking.

The hierarchical control functions of the seam tracker are implemented at the highest level by three main modules: the Supervisory Control Module (SCM), the Robot Control Module (RCM) and the Range image Processing Module (RPM). The SCM is responsible for a wide range of functions such as, system initialization, operation of the laser ranging sensor, both high-level and low-level control of the system, and communication with the RCM. The raw range data from the LPG is processed within the RPM to identify the seam's features and compute the coordinates of its characteristic features. The RCM, upon receiving the control input from the SCM, computes the required joint angles of the robot arm. While the SCM and the RPM are hosted in a microcomputer, the RCM resides in the Robot Controller.

This chapter provides an overview of the seam tracking system from the perspective of implementing a hierarchical control structure using off-the-shelf hardware components. The subsystems comprising the integrated system and the operation of the seam tracker are also discussed in detail. The concurrency requirements for the operation of such an integrated system are presented using a petrinet model.

6.2 SYSTEM DESCRIPTION

The important subsystems comprising the seam tracker include the laser range sensor, the robot manipulator and its controller, and the supervisory microcomputer. The architecture of such a system is shown in Figure 6.1. Although the SCM and the RPM are hosted in the same microcomputer, this architecture allows the RPM tasks to be easily ported to a separate processor for improving system performance. The following sections present the details of each subsystem and its operation.

6.2.1 Scanning Beam Laser Range Sensor

The laser range sensor is a Laser Profiling Gage (LPG), developed by Chesapeake Laser Systems, Lanham, Maryland, USA. It interfaces with the microcomputer through the VME databus. While the LPG is the actual sensor, the interface serves as a medium for transmitting commands to the LPG and receiving the sensor's output and status.

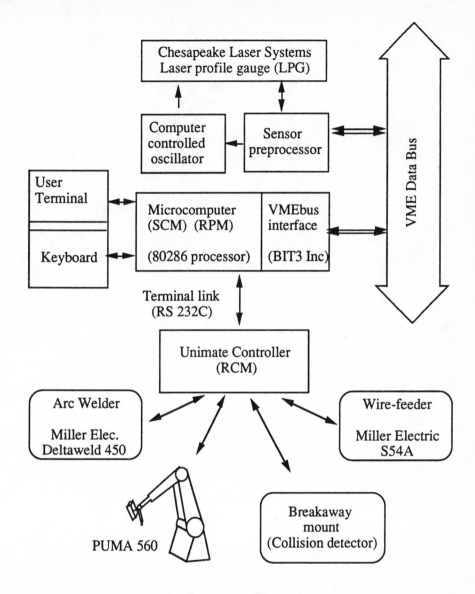

Figure 6.1 Architecture of the seam tracking system.

6.2.1.1 LPG Features. The profiling gage illuminates the surface by serially projecting a single beam of light, which upon reflection is sensed by a 1024 element, off-axis, linear CCD array. The LPG uses the principle

133

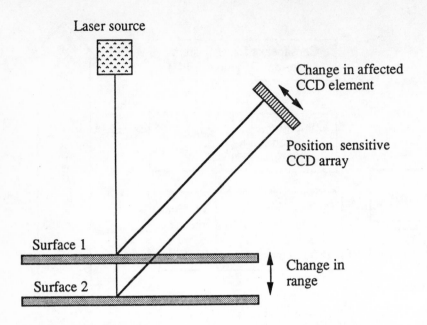

Figure 6.2 LPG's operating principle based on laser triangulation.

of laser triangulation to derive the range to the target Lambertian surface. Figure 6.2 illustrates the operating principle of the LPG.

This laser profiling gage has two distinct advantages over other serial scanning range sensors. First, the beam position is programmable and the scanning is accomplished with no moving parts (an acousto-optic deflector bends the beam through the appropriate scan angle, rather than a rotating mirror). As a result, the beam motion is instantaneous with the acousto-optic deflector being excited by radio-frequency oscillations from a computer-controlled oscillator (CCO). To ensure that the beam remains parallel to the nominal plane of the scanner and that the laser beam's diameter is 0.008 ± 0.003 inches at the focal length, a telecentric scan lens is used to focus the beam. The second advantageous feature is the programmable exposure of the CCD array, which allows the LPG to function in a wide range of light intensities and on a variety of surfaces. The controllable exposure time for the CCD array at each point along the scan allows the LPG to adapt to changes in the surface reflectance. This feature is not supported by structured light systems since the illumination of the surface is accomplished with a planar sheet of light.

These advantages make the Laser Profiling Gage extremely versatile, robust, fast and accurate.

6.2.1.2 LPG Description. The laser beam is emitted from a 30 milliwatt Gallium-Arsenide (GaAs) solid state laser diode in the near infrared region (wavelength = 780 nanometers). The beam power is selectable between 1 milliwatt and 30 milliwatts. Other specifications of the LPG include:

Accuracy: ± 0.005 in. at a nominal standoff distance of 10.0 in;
Measurement Range: 1.5 in. ± 0.75 in;
Scan Width: 1.25 in;
Beam Positioning Resolution: 26 microns.

6.2.1.3 LPG Exposure Control. The operation of the LPG consists of setting the operational parameters (CCD exposure and beam position) and acquiring the range image. Intelligent exposure control is very important since exposure time contributes directly to the control-cycle period. Proper exposure control should prevent unusually high exposure values and also prevent repeated exposures of the CCD array. Since laser triangulation is based on lambertian reflectance, it is adversely affected by reflectance extremes of two types–dark surfaces and specular surfaces (Nitzan 1987; Juds 1988). A dark surface reflects few photons and therefore, a longer exposure time is necessary to improve the signal-to-noise ratio.

To prevent long exposure times from adversely affecting the realtime requirements of the seam tracking system, the cumulative exposure count (*Cum.Expo.*) of the CCD detector and the peak CCD signal (*peak*) is monitored while recording the raw range value (*ppv*). A heuristic rule of the nature

$$If\ (Cum.Expo.\ > Threshold)\ and\ (peak < 6)\ Then\ ppv = 0$$

is applied, which if returned true indicates missing data and can be handled accordingly during range image processing. On the other hand, the time required for data acquisition of a target point is given by the equation

$$sensing\ time = exposure\ time + CCD\ transfer\ time \qquad (6.1)$$
$$= (exposure\ count \cdot 16 + 550)\ microsecs.$$

An unsatisfactory peak signal means repeating the exposure cycle, and as a consequence increasing the sensing time. An intelligent technique is therefore required to identify the correct exposure value. One such technique to compute the next exposure count employs knowledge of (i) the surface geometry, (ii) the exposure values at the corresponding target points during the previous scan, and (iii) the peak CCD signal generated from the current exposure level.

6.2.1.4 LPG Output Preprocessing. The analog output of the CCD array is preprocessed within the scanner itself. This operation consists of initial lowpass filtering followed by determination of the location of the image's centroid on the linear CCD array. The centroid location is computed from a digital representation of the image. For reliable analog to digital conversion of the analog video signals, the LPG calibration procedure is necessary. Further details on this procedure are provided in Appendix C at the end of this chapter.

The output of the LPG preprocessor includes (i) the peak CCD signal value, and (ii) the position of the affected CCD element represented by the internal parameter *ppv*. The *ppv* is a function of the range and the position of the beam in the scan. A strong CCD signal is represented by a peak value between 9 and 15 (9-F hex). The exposure level is adjusted to obtain a peak CCD signal of 10 ± 1 ($A \pm 1$ hex). Should the target surface fall outside the range of the sensor (shadow region), the output is *ppv=0* for all exposure values. The details regarding the conversion of *ppv* to range value are discussed in Appendix B at the end of Chapter 4.

6.2.1.5 Microcomputer-VMEbus Interface. The microcomputer-VMEbus interface is used to command the operation of the LPG and to transfer LPG output back to the RPM. The interface, manufactured by BIT3 Inc., consists of two printed circuit cards and an EMI shielded I/O cable to connect them. One card fits inside the microcomputer chassis and the other in the first slot of a four-slot VME bus card cage, manufactured by Ironics Inc. The address mapping of the VME bus allows the microcomputer processor direct access to memory addresses in the VME databus, just as if the memory were resident in the microcomputer. The VME bus is configured for 128K address range starting at address 1 Mbyte while the corresponding remote bus RAM space in the microcomputer is set from address hex 080000 to hex 0A0000. In this configuration, the VME adapter is designated as the bus arbiter for the VME system.

136

6.2.2 Robot Manipulator and Controller

The actuator for positioning and orienting the torch tip and the sensor along the seam is a six-axis, articulated joint, PUMA series 560 robot manipulator, manufactured by Unimation Inc., Danbury, Connecticut. The following sections discuss some its features and methods of integrating the robot into the seam tracking system.

6.2.2.1 Description. The system software that controls the robot manipulator is called VAL and is stored in the computer EPROMs located in the robot controller. The robot control module transmits instructions from the controller memory to the manipulator arm, while position data obtained from incremental optical encoders in the manipulator arm is transmitted back to the control module, thus providing a closed-loop control of the arm's motion. The PUMA robot can be programmed to interact with its environment through external input and output signals. External input signals (WX) can be used to halt the program, branch to another program step, or branch to another subroutine. The seam tracker uses this feature to halt the system in the case of torch collision, by monitoring the signal from the breakaway mounting located at the endeffector. On the other hand, external output signals (OX) allow the PUMA system to control auxiliary equipment in its work environment, including the welding controller and the wire feeder. The WX/OX features require the use of the I/O module housed in the controller, in conjunction with a relay set.

6.2.2.2 Integration with Seam Tracker. The integration of the PUMA robot within the seam tracking system involves the communication of commands and data from the supervisory control module (SCM) to the robot controller (RCM), and the communication back of PUMA's system status. An interface is provided in the robot controller to interact with an external supervisory controller for realtime path control. Upon execution of this feature, called External Alter Mode, the robot controller queries the external computer for path control information 36 times per second. The external computer must respond by sending data, which determines how the nominal robot tool trajectory is modified. However, the implementation of this scheme in the seam tracker has two basic disadvantages. First, the external computer has to respond to the robot controller every 28 milliseconds, therefore necessitating the use of a dedicated processor to handle all communication, if the main processor is

137

to remain free for supervisory control. Second, the requirement of a nominal path to be stored in the robot controller defeats the purpose of non-preprogrammed, realtime seam tracking.

For these reasons, realtime path control is accomplished using an entirely different approach that involves supervisory control from the system terminal. Normally, an operator executes VAL programs (stored in the memory) from the system terminal. Any data required by the program is input through the keyboard, while the output is displayed on the system monitor. Instead, if the VAL terminal port is interfaced with the serial I/O port of the supervisory microcomputer, the SCM can emulate the operator's functions, but at a faster speed. The robot's tool trajectory can now be generated in realtime by transmitting individual trajectory locations, one location at a time. On receiving the next pathpoint, the robot controller computes the necessary joint angles. The joint servo controller asynchronously executes these angles, and the robot controller is free to concurrently query the SCM for the next path point.

Remark: Some robot controllers feature a direct memory access scheme that allows the SCM to asynchronously transfer the path locations to the RCM memory. This feature prevents the present time consuming polling method necessary to synchronize the operation between the SCM and the RCM. ♣

Communication between the RCM and the SCM is based on the simple concept of transmitting data and receiving an acknowledgment (ACK). All communication errors are represented by negative acknowledgment (NAK), requiring retransmission of the data. Communication is handled by background processes, executing asynchronously on both the robot controller and the supervisory controller.

6.2.3 Supervisory Microcomputer

The Supervisory Control Module (SCM) is hosted in a IBM PC/AT compatible microcomputer with a 12 MHz Intel 80286-based architecture. The microcomputer is equipped with 100 nanosecond dynamic RAM along with 8 MHz DMA, counter/timer interrupt controller, clock generator chips which support speeds up to 16 MHz, and an Intel 80287 math coprocessor for expediting floating point computations. An asynchronous serial I/O communications card with two serial ports, supporting baud rates of up to 19200, is also fitted in the microcomputer. This provides the interface between the SCM and the RCM

138

and is operated at 1200 baud, with a seven bit data word, no parity and one stop bit. The buffer size for this interface is 4K, both for transmission and receiving. The microcomputer interfaces through an adapter card with the VME databus, to interact with the Laser Profiling Gage (LPG) system. The microcomputer configuration has a main memory of 1M, of which 512K is used by the operating system and other application programs while the remaining 512K is for the remote bus RAM addresses in the VME system.

The software modules in the SCM have been developed in the C programming language, reputed for its portability, performance and bit-level operations. The SCM has a modular structure, with separate modules for initializing the communication ports and the LPG, operating the LPG, invoking the range image processor (RPM), executing various other high-level and low-level control tasks, communicating with the RCM via the serial I/O port over RS 232C, and error handling.

6.3 SYSTEM CONTROL SOFTWARE

The architecture of the seam tracker is illustrated in Figure 6.1 while the hierarchical structure of the control software is shown in Figure 6.3(a) through 6.3(f). The seam tracking operations are supported at the highest level, by three main modules—Supervisory Control Module (SCM), the Robot Control Module (RCM) and the Range image Processing Module (RPM). While the SCM and the RPM are hosted in a microcomputer, the RCM resides in the Robot Controller. The operation of the seam tracking system, with its modularly structured software, involves interaction among these individual modules. The functions and structure of these modules are as follows.

6.3.1 Supervisory Control Module (SCM)

As shown in Figure 6.3(a), the SCM is responsible for various tasks including, (i) system initialization (refer Figure 6.3(b)), (ii) operation of the range sensor (refer Figure 6.3(c)), (iii) invoking RPM for seam recognition, (iv) robot endeffector trajectory control (refer Figure 6.3(d)), and (v) communication with the RCM.

To obtain reliable data, the SCM controls the range sensor's operating parameters (refer Figure 6.3(c)), which include (i) exposure time of the CCD array, (ii) beam position in the scan, and (iii) threshold levels for CCD signal preprocessing. The raw range data (ppv) so acquired is preprocessed in the RPM by converting it into distance in the range image.

139

ADAPTIVE, REAL-TIME, INTELLIGENT
SEAM TRACKING SYSTEM

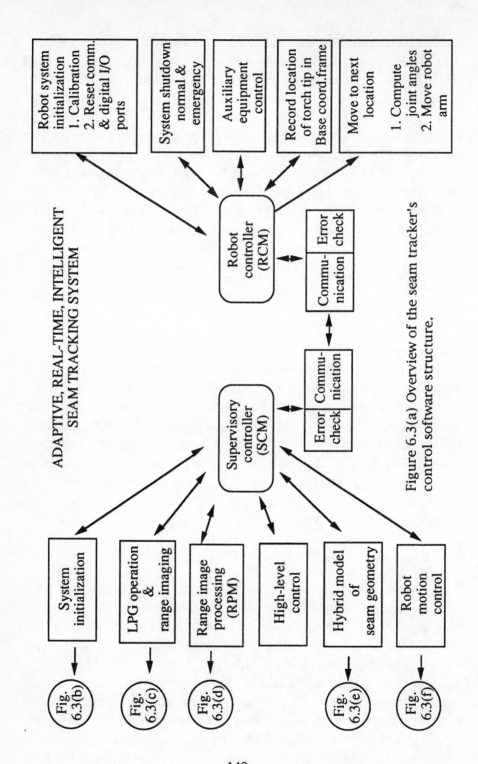

Figure 6.3(a) Overview of the seam tracker's
control software structure.

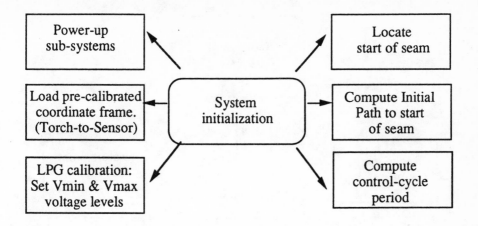

Figure 6.3(b) Control software structure for system initialization.

This is further processed by the RPM to extract the feature points (segmented range image) and compute their coordinates in world space (refer Figure 6.3(d)).

The unstructuredness of the seam and its environment requires adaptive modeling at two levels—modeling the seam environment and the seam geometry. The seam environment is modeled as a sequence of seam types (vee, butt, or fillet joints, tack welds, end-of-seam condition or torch runaway condition), corresponding to each scan along with other

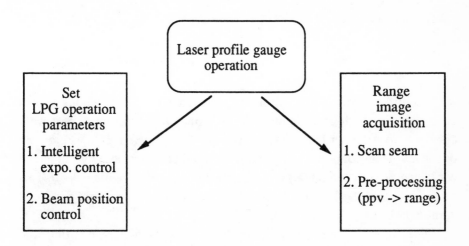

Figure 6.3(c) Control software for laser profiling gage operation.

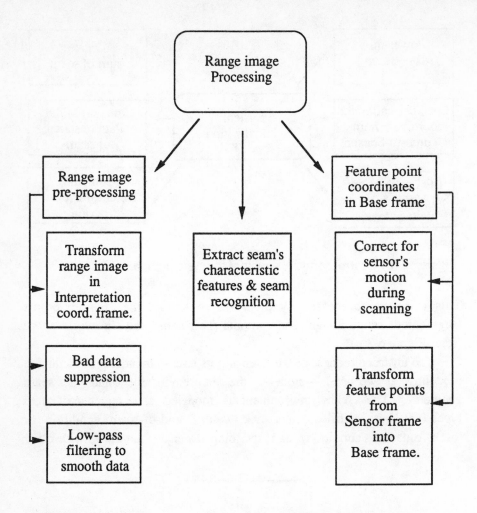

Figure 6.3(d) Control software structure for range image processing.

related information. This information can be used by the high-level controller for torch-seam interaction in conjunction with an expert weld planner. On the other hand, the locations of the extracted features are used by the low-level controller to represent the seam geometry as a hybrid model consisting of both continuous and discrete models (refer Figure 6.3(e)). This seam model provides the basis for the control input required for robot motion control.

The input to the robot controller consists of (i) position and orientation of the torch along the seam, and (ii) orientation of the sensor over the seam (refer Figure 6.3(f)). All interprocess communication

142

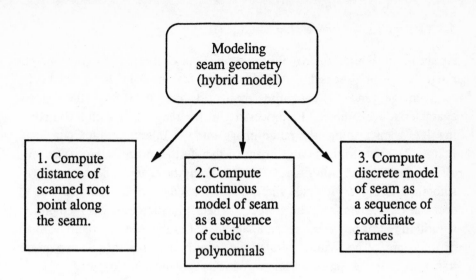

Figure 6.3(e) Control software structure for computing hybrid model of
the seam geometry.

between the SCM and RCM is carried across an RS232C asynchronous
serial I/O channel.

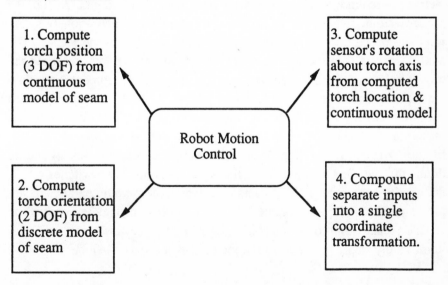

Figure 6.3(f) Control software structure for endeffector trajectory control.

6.3.2 Range Image Processing Module (RPM)

As shown in Figure 6.3(d), the main functions of this module include, (1) range image preprocessing, (ii) extracting feature point locations in the range image, and (iii) computing the world coordinates of the seam's characteristic features. Preprocessing of the range data within the RPM involves transforming the range image into the Interpretation Coordinate Frame. The transformed range image is then further processed to (i) remove all bad data due to multiple reflections from the scanned seam, and (ii) smoothed, using a lowpass filter, to alleviate the effect of noise due to specular reflections from the surface. The application of seam recognition algorithm to this processed range image identifies the characteristic features of the seam. The feature points thus identified in the range image are transformed in the Base coordinate frame as explained in Chapter 5.

6.3.3 Robot Control Module (RCM)

The RCM, upon receiving the control input from the SCM (in cartesian coordinates), computes the required joint angles using the inverse kinematic solution of the robot arm. As each joint moves through its angle, the welding torch moves along the seam with the correct orientation and the proper welding speed. The joint motion is controlled asynchronously by the joint servo-controller working at higher frequencies than the RCM. The control input also prescribes the required rotation of the sensor about the torch's axis to track the seam ahead. The auxiliary equipment, including the welding controller (Miller Electric Deltaweld 450) and the wirefeed controller (Miller Electric S54A) interface with the robot controller in a master-slave structure, with the robot controller as the master (refer to Figure 6.1). The welding controller receives commands for switching on/off the welding current, while the wirefeed controller switches on/off the wirefeed mechanism. In this implementation, the welding parameters are preset on the welding equipment itself. A collision detector within the torch's breakaway mount interfaces directly with the robot controller for immediately stopping the robot motion in case of torch collision.

6.4 SYSTEM OPERATION

The operation of the seam tracker can be understood in terms of (i) the initial system calibration, (ii) system initialization during powerup, and (iii) the normal seam tracking sequence. Since SCM and the RCM are hosted in two separate subsystems (i.e., the microcomputer and the robot controller), it is also important to understand the synchronization between these two concurrently operating modules during realtime seam tracking.

6.4.1 Initial System Calibration

The various coordinate frames required for realtime torch path generation are defined once for each workcell during the calibration process. They do not have to be redefined unless there is a change in the system or the work environment.

As discussed in Chapter 4, realtime path generation requires the use of four coordinate frames, namely,

1. Base coordinate frame;
2. Torchtip coordinate frame;
3. Sensor coordinate frame;
4. Seam coordinate frame.

The first three coordinate frames have to be predefined during the calibration process. The Seam coordinate frame, on the other hand, is computed from the range image during each seam tracking cycle. The Base coordinate frame is predefined at the base of the robot and is the global reference frame for the system. The other two coordinate frames, i.e., the Torchtip coordinate frame and the Sensor coordinate frame, are defined as per the calibration process described in Chapter 4.

Although the Sensor frame is defined relative to the Base frame, it is helpful to calculate it relative to the Torchtip coordinate frame as well. This coordinate transformation, *Torch-to-Sensor* $[T_2S]$ is required while transforming the feature point coordinates from the image space into the world space using known torchtip locations.

6.4.2 System Initialization during power-up.

During this step, all subsystems, including the robot controller and manipulator, welding controller, wirefeeder, microcomputer, and the LPG, are powered up. The robot arm is calibrated to locate the zero positions of the joints. The VAL operating system and various application programs are uploaded into the robot controller EPROMs from floppy disks. This also includes previously defined transformations relating the various coordinate frames. The LPG is calibrated to set the voltage levels V_{min} and V_{max}

necessary for its operation. The operational parameters of the LPG have to be calibrated everytime there is a change in the ambient light, temperature, and the overall reflectivity of the part's surface. Appendix C discusses the procedure for calibrating the LPG.

Immediately following the power-up phase, the initialization procedure in invoked as shown in Figure 6.3(b), This includes,

(i)   SCM receiving the current values of the Torch-to-Sensor coordinate frame from the RCM;
(ii) setting the LPG's operational parameters;
(iii) computing the control-cycle period (t_{cycle}).

6.4.2.1 Computing Control-Cycle Period (t_{cycle}). This represents the cumulative time required for:

1. scanning the seam / finding the current location of the torchtip by sensing the robot arm joint encoders;
2. communicating the torchtip location, at the start of scan, to the SCM;
3. range image processing;
4. generating models of the seam's environment and geometry;
5. computing the control input for positioning the torch and the sensor;
6. communicating the control input to the Robot Controller.
7. computing the inverse kinematic solution of the robot arm joint angles in the RCM;
8. additional buffer time.

It should be noted that the control-cycle period is constant, since all the above processing steps are required during each cycle. The above sequence of events provides the initial estimate of the control-cycle period $(\hat{t}_{cycle})$. This is further refined during computation of the Initial Path as described in Section 6.4.3.

Assuming a constant tracking speed during any cycle, the control-cycle period (t_{cycle}) is directly related to the distance travelled by the torch per cycle (d_{cycle}) as given by

$$d_{cycle} = t_{cycle} \cdot V_{weld} \qquad (6.2)$$

From the standpoint of seam tracking accuracy, a shorter d_{cycle} is preferred for two reasons: First, although the seam being tracked may be nonlinear, the motion of the torch between locations along the seam is piecewise linear. Hence, smaller chords along the seam will introduce smaller errors while tracking nonlinear seams. Second, since the scans can now be closely spaced, a more accurate model of the seam can be generated.

6.4.3 Seam Tracking Sequence

The complete process of seam tracking during robotic welding can be partitioned into three phases: initial path tracking, normal seam tracking and shutdown. These phases are described in the following sections.

6.4.3.1 Initial Path Tracking. To start the seam tracking process, the operator positions the torch tip with the desired orientation at the start of the seam given by location $[R_I]$. Alternately, the start of the seam can be reached automatically based on CAD information of the part and using part finding algorithms (Nayak et al. 1990; Agapakis et al. 1990b). Note that at this location, the sensor is not at the seam start. To locate the seam start, the *Torch-to-Sensor* transformation $[T_2 S]$ is used to move back the Sensor Frame's origin to the current position of the torch, while maintaining the proper standoff distance. This step may not be necessary when the torch-to-scanner distance is short since the seam in between can be assumed to be straight. However, in our implementation the relatively large torch-to-scanner distance necessitated moving back the scanner to the seam start.

After scanning the surface, the seam is located and the system correctly positions the sensor's scanning window over the seam. This new position of the torch, which is outside the seam, is given by

$$[R(-n)] = [R_I][T_2 S]^{-1} \qquad (6.3)$$

and the vector ($D_{initial\ path}$) relates the two torch locations $[R(-n)]$ and $[R_I]$.

Refining control-cycle period estimate ($\hat{t}_{cycle}$). The total time $T_{initial\ path}$ required to traverse the initial path vector $D_{initial\ path}$ along a straight line trajectory, at a preset welding speed V_{weld} is given by

147

$$T_{initial\ path} = \frac{D_{initial\ path}}{V_{weld}} \qquad (6.4)$$

Based on the initial estimate of the control cycle period $(\hat{t}_{cycle})$, the initial path can now be divided into $N_{initial\ path}$ number of integral segments, using the relation

$$N_{initial\ path} = \text{Mod}\left(\frac{T_{initial\ path}}{\hat{t}_{cycle}}\right) \qquad (6.5)$$

The distance traveled per cycle (d_{cycle}) is then given by

$$d_{cycle} = \frac{D_{initial\ path}}{N_{initial\ path}} \qquad (6.6)$$

For a preset tracking speed, the final value of the control cycle time is given by

$$t_{cycle} = \frac{d_{cycle}}{V_{weld}} \qquad (6.7)$$

The number of path locations in the initial path, $N_{initial\ path}$, is also defined as the *torch-lag* (n), i.e., the number of steps the torch lags behind the lookahead sensor.

The location of the $N_{initial\ path}$ path points along the Initial Path are computed using a drive transformation relating $[R(-n)]$ to $[R_I]$ as discussed in Chapter 5. The motion along the Initial Path ends with the torch positioned at the seam start, whereupon the shielding gas and the welding arc are switched on after the required predwell period. This is followed by the normal seam tracking cycles until shutdown.

6.4.3.2 Normal Seam Tracking. The motion of the torch during normal seam tracking is based on the adaptively generated model of the seam geometry. The sequence of operations executed during each cycle are illustrated in Figure 6.4. As shown in this figure, there are four concurrent operation cycles in the system: (i) the robot arm motion, (ii) the RCM operation, (iii) the SCM operation, and (iv) the range sensor operation. At the start of each cycle, data is acquired from two different sensors—the RCM senses the joint encoders for computing the current torchtip location

148

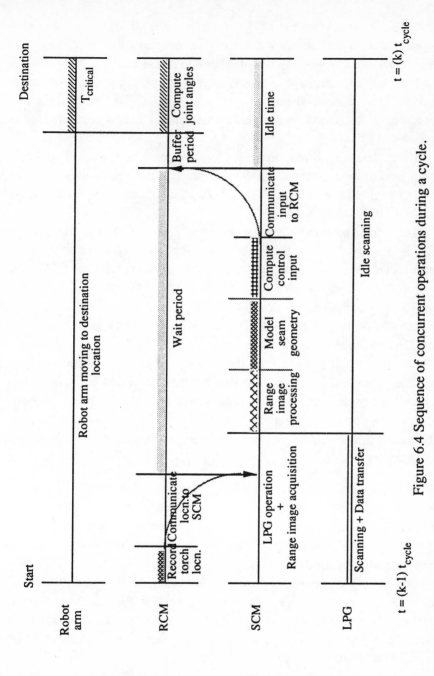

Figure 6.4 Sequence of concurrent operations during a cycle.

149

and the SCM commands the LPG to acquire a range image. Thus, data acquisition from both sensors is carried out concurrently. Normally, the joint encoder data acquisition and its communication to the SCM takes only a fraction of the time required for range imaging the seam. During the range imaging process, the robot arm is in motion. The effect of the arm movement on the world coordinates of the feature points is accounted for as discussed in Chapter 5.

After data acquisition, the sequence of range image processing, seam model generation, and computation of the control input follows in the SCM. During this period, the RCM goes in the waiting mode for receiving data from the SCM.

The time for communicating the control input from the SCM to the RCM is directly added to the cycle time and hence, any retransmission is undesirable. Upon receiving the next pathpoint (k+1) from the SCM, the RCM computes the required joint angles during the period $t_{critical}$ before reaching the current destination (k). During this period, the SCM idles till the start of the next cycle. When the torch reaches its current destination location, a new cycle starts.

In the present architecture, the major time expenses incurred are in range imaging and in communication between the RCM and the SCM. Any reduction in these times would make significant contribution towards reducing the control-cycle period. This is a major reason for preferring robot controllers with direct memory access since they have small communication delay.

6.4.3.3 Termination of Seam Tracking. The high-level controller governs the torch-seam interaction based on the seam environment model. When the end-of-seam condition, or an unknown seam condition is encountered, the system tracks the last segment of the seam and also starts preparation for terminating the seam tracking operation. This includes halting the scanning during each cycle and informing the RCM when the last location along the seam is reached. On reaching the last location, the welding arc and the shielding gas are switched off after the required postdwell period, the wire feeder is switched off, and the torch tip is moved away from the seam.

Most emergency shutdown situations are caused by either activation of the collision detector or by malfunctioning of the auxiliary equipment such as the wirefeeder, welding controller, the shielding gas solenoid valves, or the water cooler. Since the auxiliary equipment is controlled by the robot controller, it also initiates the emergency shutdown procedure. This includes stopping the robot arm motion and shutting down all auxiliary equipment. The RCM then informs the SCM of the emergency situation to effect a complete shutdown of the entire system. Restarting the system is simple and requires placing the torch at the start of the remaining section of the seam to be tracked.

6.5 SYNCHRONIZING CONCURRENT OPERATIONS IN SEAM TRACKING

The seam tracker features concurrent operation of the SCM, the RCM, and the robot arm. An important consideration is the proper synchronization of these concurrent operations to ensure smooth tracking of the seam. This requirement is modeled through the Timed Petrinet (Peterson 1981) in Figure 6.5, which illustrates the operations of the RCM, the SCM and the robot arm along with the interprocess communication involved. The reader if not familiar with petrinet concepts and terminology, may proceed to the next section without loss of continuity.

During initial startup, the tokens in places s_1, s_2 and s_{10} indicate that all three subsystems, i.e., the SCM, the RCM, and the robot arm, respectively, are ready to execute. Firing of transition e_1 follows, indicating that the robot controller has received the control input to move the arm to the next destination location. The integrity of this data is checked in the RCM (s_3) while the SCM waits for acknowledgment from the RCM (s_4). If no errors are detected (e_2), then an ACK is transmitted to the SCM (s_6) indicating successful communication. The SCM can now proceed with subsequent operations in its cycle. However, if an error is detected (e_2) then a negative acknowledgment (NAK) is transmitted to the SCM (s_5) and both the RCM and the SCM return to their initial states (s_1, s_2) for retransmission (e_3). Eventually, after successful data communication (e_4), the RCM receives the next destination location (s_8). If this is the first location in seam tracking, as indicated by a token in s_{10}, then the transition e_8 fires and the joint angles for the robot arm's motion are computed (s_{13}) with subsequent motion to the destination location (s_9). The bottomline in the entire seam tracking operation is to ensure smooth motion of the torch tip from one location to the next without stopping. This is possible only if the next destination location is available to the RCM at the instant $\left(t_{cycle} - t_{critical} \right)$ during every cycle (also refer to Figure 6.4). Tokens in s_{11} and s_8 represent this situation, which ensures smooth realtime path

151

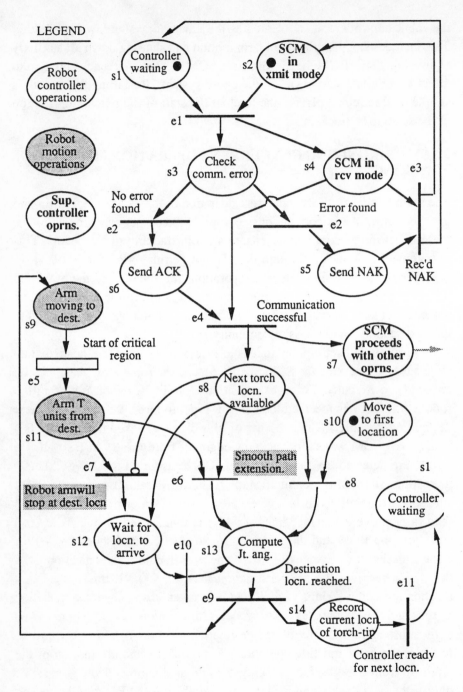

Figure 6.5 Petrinet model of realtime robot path motion for seam tracking.

152

generation (e_6). The firing of the timed-transition e_5 indicates the start of the *critical period*. During this time period, the RCM computes the necessary joint angles for the robot arm to reach the next destination while maintaining the proper velocity profile. A situation wherein s_{11} has a token but none is present in s_8 represents the arrival of an input location at the RCM after the start of the *critical period*. This results in deceleration of the arm to zero velocity after the current destination is reached (e_7). Later, upon receiving the destination location, the robot arm does resume tracking. However, such stoppages before moving to the next destination are highly undesirable.

As seen in the Petrinet, the stopping of the robot arm is directly affected by the communication between the RCM and the SCM; any retransmission only increases the possibility of pushing communication into the critical zone. The Petrinet, thus provides an understanding of the critical operations involved in synchronizing the concurrent processes of the SCM, the RCM, and the motion of the robot arm, which can then be made more efficient.

6.6 SUMMARY

The architecture of the seam tracking system comprises a microcomputer, a laser range sensor, a robot controller with a six DOF manipulator, and the auxiliary equipment for welding and safety considerations. The seam tracker's hierarchical control software is structured in three modules: (1) the supervisory control module (SCM), (2) the robot control module (RCM), and (3) the range image processing module (RPM). In the development of the seam tracking system with its distributed subsystems, it is important to understand the interaction amongst the concurrent operations of these subsystems. This has been accomplished through a petrinet model of the system's critical operations.

The following chapter analyses the limitations of seam tracking systems and the sources of various seam tracking errors. Specifications are also drawn up for the minimum radius of curvature of the seam based on various criteria.

APPENDIX C

C.1 LASER PROFILING GAGE CALIBRATION

The calibration of the Laser Profiling Gage is required to set correct thresholds on the light and dark levels of the CCD array and thus, provide a proper contrast for the video analog to digital conversion. By adjusting the dark-level voltage (V_{max}), we attempt to compensate for the ambient lighting, optical noise and temperature fluctuations. As ambient lighting intensity increases, the V_{max} voltage goes down. Also, as the ambient temperature increases, the V_{max} voltage decreases. A decreasing dark-level reduces the contrast of the video image. The adjustment of the light-level voltage (V_{min}) is required if the integration time of the CCD array varies. The V_{min} voltage is set just above the saturation level of the CCD array. Both voltage levels should be reset whenever there is a drastic change in the lighting condition.

Figure C.1 shows the setting of the V_{max} for calibrating the ambient light level while Figure C.2 shows the setting of the V_{min} voltage level. The calibration process (also called *autodac*) has the following sequence:

1. Set the light-level voltage (V_{min}) to a low value.
2. Set the dark-level voltage (V_{max}) to a high value.

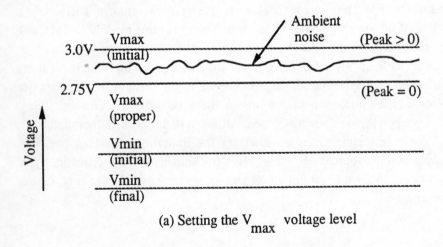

(a) Setting the V_{max} voltage level

Figure C.1 Setting V_{max} voltage to calibrate ambient light level.

154

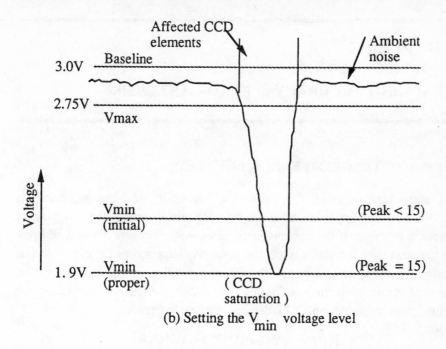

(b) Setting the V_{min} voltage level

Figure C.2 Setting V_{min} voltage levels for given surface reflectance.

3. Position the laser spot out of the field of view of the receiver CCD array by setting CCO to an invalid low value (CCO=0). This leaves a noise baseline of the surface.

4. Reduce V_{max} until the peak CCD signal goes to 0. Reduce V_{max} byte roughly 6 more counts, just to be sure.

5. Position laser beam to fall on target (CCO=9501).

6. Adjust exposure time to make sure the CCD array gets saturated.

7. Vary V_{min} until the peak CCD signal = F (hex).

8. Reset to desired exposure and beam position.

CHAPTER 7

TRACKING ACCURACY & ERROR ANALYSIS

7.1 IS THE TRACKING ERROR BOUNDED?

An important criterion for validating the seam tracking algorithm is establishing an upper bound on the tracking error during the entire seam tracking process. If the tracking error does not propagate from one cycle to the next then the stability of the seam tracking process is ensured. The pathpoint coordinates input to the robot controller during each cycle is derived from both the continuous model and the discrete model of the seam, given by Equation (7.1) derived earlier in Chapter 5.

$$[U(k)] = [Y(k+1)][Z(k+1)][C_2R] \qquad (7.1)$$

The position and orientation of the torch is part of the control input $[U(k)]$ and this component is given by $[Y(k+1)]$ as described by Equation (7.2).

$$[Y(k+1)] = \begin{bmatrix} n_x(k+1) & o_x(k+1) & a_x(k+1) & \bar{p}_x(k+1) \\ n_y(k+1) & o_y(k+1) & a_y(k+1) & \bar{p}_y(k+1) \\ n_z(k+1) & o_y(k+1) & a_y(k+1) & \bar{p}_z(k+1) \\ 0 & 0 & 0 & 1 \end{bmatrix} \qquad (7.2)$$

Since $[Y(k+1)]$ is a function of the computed root position $\bar{p}(k+1)$ and the vectors $[n, o, a,]$, any errors in these computed values would affect the position and orientation of the torch along the seam. For welding applications, the vectors $[n, o, a,]$, representing the torch orientation, have a fairly large allowable tolerance and so any deviations in the torch orientation within tolerance, will not adversely affect the welding quality. However, the torch position along the seam is very critical to maintaining good weld quality. The torch position is derived from the feature point coordinates $w_2(k)$, given by Equation (7.3).

$$\{w_j(k)\} = [R_{i(j)}(k-n)][T_2S]\{Q_j(k)\} \quad j=1...3, \quad i(j)=0...r-1 \qquad (7.3)$$

where,

$\{Q_j(k)\}$ are the jth feature point coordinates in image space;

$[T_2S]$ is the fixed transformation relating the Torchtip and Sensor frame;

$[R_{i(j)}(k-n)]$ is the location of the torchtip frame while scanning the jth feature point or the $i(j)^{th}$ scan point.

From this relation, to correctly compute the feature point coordinates in world space, the position of the robot endeffector $[R_{i(j)}(k-n)]$ should be correctly recorded during every scan $\{S_i(k)\}$. This approach ensures a stable seam model because the endeffector location $[R_{i(j)}(k-n)]$ is actually recorded and not based on pathpoint coordinates computed earlier. Since the actual torch position is recorded at each step, any error in postioning the torch over the seam at the $(k-n)^{th}$ pathpoint is not incorporated in the kth feature point $w_2(k)$ and subsequently, the kth pathpoint coordinates $[Y(k+1)]$. In effect, this approach allows one to generate a correct world model of the seam even with the torch positioned completely outside the seam.

Since any error in seam tracking during the current cycle does not affect the path point coordinates computed in the future cycles, the seam tracking error is essentially bounded. Its maximum value during any cycle establishes the upper bound on the tracking error for the entire seam tracking process.

The following sections analyze the various sources contributing to the seam tracking error during any cycle. To keep the seam tracking error within specified bounds, it is also helpful to know the limitations of the seam tracker. This is presented in the form of specifications on the minimum radius of curvature allowable for any seam.

7.2 SOURCES OF ERROR

During any seam tracking cycle, the tracking error can be attributed to a variety of factors, which are broadly categorized as:

1.   Error in computing the pathpoint coordinates (control input);
2.   Error in tracking the pathpoint coordinates.

The following sections explain the causes of the various errors although the significance of these errors on seam tracking accuracy is not discussed. The detailed quantitative analysis required for such a task is outside the scope of this monograph.

7.2.1 Error in Computing the Pathpoint Coordinates

An incorrect pathpoint coordinate value could be computed as a result of:

7.2.1.1 End-of-Arm Tooling Errors. These include the errors in defining the various coordinate frames associated with the end-of-arm tooling such as the torch and the lookahead sensor mounted at the endeffector. For robotic welding application, the two coordinate frames defined for this purpose include, the Torchtip coordinate frame [T] and the Sensor coordinate frame [S]. If the origin of the Torchtip coordinate frame does not coincide with the torchtip, then (i) a rotation about any-axis of the torchtip frame would result in a position error at the torchtip, and (ii) the pathpoint coordinates will suffer from a constant bias error. These errors can be reduced by careful definition of the coordinate frames.

7.2.1.2 Sensing Errors. All sensors are resolution limited. If accuracy were to be improved, the resolution of the measurement devices has to be better than the required accuracy. The two kinds of sensors in the system are the encoders in the robot arm, and the ranging sensor. Seam tracking accuracy is affected by the resolution of both these sensors. Environmental factors also affect the accuracy of the measurement devices. Variations in the ambient temperature can cause the sensor electronic parameters to drift from their nominal values affecting both the robot accuracy and range measurements. This, however, can be minimized through controlled ambient temperature.

Encoders in the robot arm can be mounted either directly to the motor shaft or the joint axes. Encoders mounted directly to the motor provide ease of motor commutation control but parametric factors such as flexibility and backlash affecting robot accuracy are not fed back to the robot controller. On the other hand, encoders mounted directly to the joint axes although more accurate, require high resolution to detect small increments of joint positions. To accurately command robot moves in the cartesian frame, a common reference should be established between the encoders and the actual joint angles. The robot calibration procedure, which establishes the zero reference positions for the robot joints, is therefore necessary (more details regarding this procedure are provided in Appendix D at the end of this chapter).

Erroneous range measurements are generated due to multiple reflections, which reduce the true range value at the scanned point in vee-grooves. Heuristics, based on this knowledge, should be used to preprocess the noisy range image.

In scanning beam range sensors, the resolution of the sampled range data across the seam section introduces errors in the seam model. To reduce the scantime, the data is sampled only at certain positions along the scan making the range image granular. As a result, the computed feature points along this granular range image do not precisely correspond to the true seam features (refer to Figure 7.1), thereby affecting the x and z coordinates of the feature points in the image space. In the context of seam tracking for robotic welding, the z–direction error (e_z) will affect the arc length (torch to seam distance) and the x–direction error (e_x) affects the tracking accuracy. The sampling frequency of structured light range sensors on the other hand, is much higher than serial beam scanners since they illuminate the entire seam section at once. However, the computational burden introduced by the large volume of data may require reduction in the sampling frequency during subsequent processing.

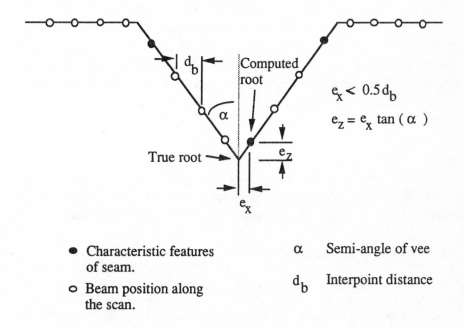

$$e_x < 0.5\, d_b$$
$$e_z = e_x \tan(\alpha)$$

- Characteristic features of seam.
- Beam position along the scan.

α Semi-angle of vee

d_b Interpoint distance

Figure 7.1 Effect of range data sampling on feature point location.

159

7.2.1.3 Robot Inaccuracy. Industrial robots move to programmed positions as directed by the robot's control equations. Control equations such as the inverse kinematics equations of robot motion are derived from a mathematical model of the robot joint and link structure. The "ideal arm" model used in the robot design normally becomes "nonideal" due to manufacturing errors. This deviation between the robot's ideal mathematical representation and the physical robot will result in robot inaccuracy.

The robot's pose $[R(k-n)]$ at the start of each seam tracking cycle is recorded by the robot controller using joint encoder values. The robot inaccuracy affects the recorded value of the torchtip position in cartesian frame $[R(k-n)]$, which subsequently affects the feature point coordinates in world space computed from $[R(k-n)]$, and eventually the final pathpoint coordinates $[U(k)]$ for seam tracking. Common causes of robot inaccuracy include structural imperfections in the link lengths, joint misalignment, gear backlash, bearing wear/wobble, encoder error, joint and link compliance, temperature variation, etc. Appendix D at the end of this chapter discusses the issues related to robot accuracy in greater detail.

7.2.1.4 Seam Modeling Errors. The true function of the seam being unknown, it is necessary to approximate it. This approximate model of the seam is also a source of errors while computing the control input. Any approximating function should behave well not only at the data points, but also in the region between the data points. In the case study presented in Chapter 5, the seam has been approximated using a sequence of least squares cubic polynomials in parametric form, computed from a moving data window of feature point coordinates. The least squares approach has the advantage of smoothing out the noisy data. Further, by choosing a small window size one can accurately model the corresponding section of the seam and as new polynomial coefficients are computed, they incorporate the changes along the seam. Consequently, the seam model can be considered stable and adequate for representing the true seam.

7.2.2 Error in Tracking the Pathpoint Coordinates

This error is attributed to the following factors:

7.2.2.1 Piecewise Linear Motion of the TorchTip. In our particular implementation of the seam tracker described in Chapter 6, the system is based on off-the-shelf components. The path generation module resides

160

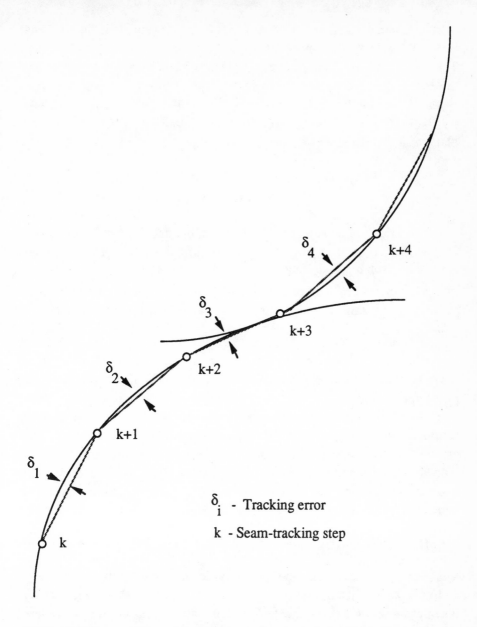

δ_i - Tracking error

k - Seam-tracking step

Figure 7.2 Error due to piecewise linear motion of the torchtip.

in the supervisory controller separate from the motion control module resident in the robot controller. With such a system, while the seam may be nonlinear, the motion of the torch from one point on the seam to the

161

next is linear. This piecewise linear travel of the torchtip along the nonlinear seam introduces an error δ_x in tracking along the x–direction of the Torchtip coordinate frame, as seen in Figure 7.2. For a given curvature of the seam, this error increases with the torch travel distance per cycle, d_{cycle}. This in turn is a function of the tracking speed V_{weld}, assuming a constant control-cycle period t_{cycle} for the system, as described by Equation (7.4).

$$\delta_x \propto d_{cycle} \qquad (7.4)$$
$$\propto V_{weld} \, t_{cycle}$$

This error can be reduced by decreasing the tracking speed, especially in sections of seam with rapidly changing geometry. The maximum allowable error δ_x is an important parameter for specifying the minimum radius of curvature that can be tracked by the system. For welding applications, maximum allowable error is normally half the wire diameter.

7.2.2.2 Robot Inaccuracy. During seam tracking, the robot moves are commanded in the cartesian frame. As discussed earlier, just as robot inaccuracy affects the recording of the torchtip location, it also affects the positioning of the torchtip at the commanded pathpoint in cartesian frame. The "nonideal" implementation of the "ideal" arm inverse kinematic solution results in the torchtip not reaching the commanded pathpoint location. This error is further compounded by any errors in coordinate frames associated with end-of-arm tooling. Although the position errors resulting from robot inaccuracy translate directly at the torchtip, all orientation errors are compounded at the torchtip by a factor proportional to the tool size.

7.3 MINIMUM RADIUS OF CURVATURE FOR THE SEAM

The errors discussed in the previous section affect the performance and accuracy of seam tracking. On the other hand, there are certain inherent limits for a seam tracking system based on its design parameters, such as the distance between the torch and the lookahead sensor, the field of view of the triangulation-based range sensor, the width of the scan or structured light plane, and the system's control-cycle period. For robust seam tracking, a threshold is set on the seam's minimum radius of curvature (R_{min}), below which the tracking error increases significantly. This section presents the preliminary investigation into specifying the minimum radius of curvature for any seam based on certain preconstructed situations.

In our analysis, the minimum radius of curvature (R_{min}) for a seam is a function of three variables:

1. the allowable tolerance on the tracking error due to piecewise linear motion of the torchtip along a nonlinear seam;
2. the field of view of the range sensor;
3. the torch-sensor assembly design parameters (preview distance L, scanwidth $2b$) and the width ($2c$) of the seam being tracked.

The specifications for the seam's minimum radius of curvature are discussed in the following sections.

7.3.1 Allowable Tracking Error (δ_{max})

The seam being tracked can be nonlinear but the motion of the torch along

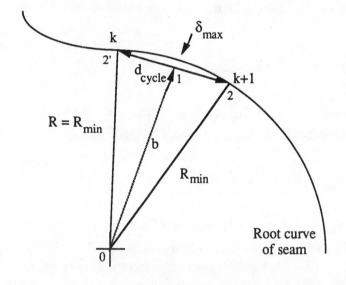

d_{cycle} : Distance travelled per cycle

R_{min} : Minimum radius of curvature

δ_{max} : Maximum allowable tracking error

Figure 7.3 Minimum Radius of Curvature based on Allowable Tracking Error (δ).

163

the seam is piecewise linear (Note that some robot controllers allow circular interpolated motion as well. But in this discussion, it is assumed that the torch will be traveling along a linear path between two computed locations). Figure 7.3 shows the tracking error (δ) introduced due to the linear motion of the torchtip. Although this error has components along all three coordinate axes of the Torchtip frame, only the components in the x and y directions affect the tracking accuracy. Hence, the minimum radius of curvature is computed in the xy-plane of the Torchtip frame.

In Figure 7.3, consider the right triangle **0-1-2**, which is generated due to the bisection of triangle **0-2'-2** by the line b. The minimum radius of curvature in this case is a function of the torch travel per cycle, given by

$$R^2_{min} = b^2 + \left(\frac{d_{cycle}}{2}\right)^2 \tag{7.5}$$

where, $b = R_{min} - \delta_{max}$.

Substituting for b in Equation (7.5), the minimum radius of curvature is given by

$$R_{min} = \frac{\delta_{max}}{2} + \frac{d^2_{cycle}}{8\delta_{max}} \tag{7.6}$$

In terms of the tracking speed V_{weld} and the control-cycle period t_{cycle} of the seam tracking operation, the minimum radius is given by

$$R_{min} = \frac{\delta_{max}}{2} + \frac{\left(V_{weld}\, t_{cycle}\right)^2}{8\,\delta_{max}} \tag{7.7}$$

This value of R_{min} is based on the projection of the actual 3D seam in the xy-plane of the Torchtip coordinate frame. Usually, the radius of curvature of the projected seam would be greater than the corresponding actual radius of curvature of the 3D seam R in any other plane. Hence, this analysis represents a limiting case where the seam lies in the xy-plane of the Torchtip coordinate frame.

7.3.2 Range Sensor's Field of View Limitation (δ_r)

One problem with range finding based on triangulation is that there is no range data in the "shadow" region (portions of the scene that are not visible

164

to both the projector and the camera). Figure 7.4 illustrates a situation where the seam being tracked is at the outer limit of the range sensor's field of view δ_r. As seen in this figure, the origins of the Torchtip frame and the Sensor frame lie along the tangent to a circular seam, where the tangent is drawn at the current position of the torchtip. In this configuration, the range sensor equally favors the downward sloping or upward sloping seam.

In Figure 7.4, consider the two triangles: isoceles triangle **0-1-2**, and right triangle **1-2-3**.

In triangle **0-1-2**,

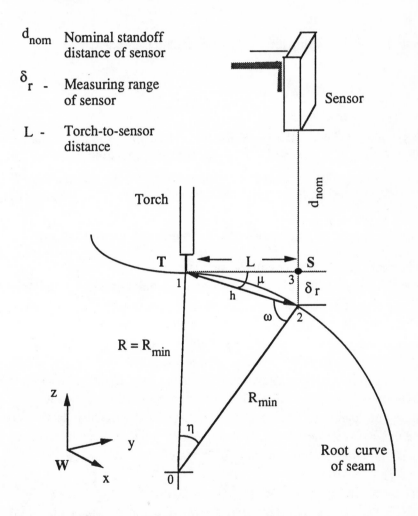

d_{nom} Nominal standoff distance of sensor

δ_r -   Measuring range of sensor

L -   Torch-to-sensor distance

Figure 7.4 Minimum radius of curvature based on measuring range of ranging sensor.

$$h^2 = 2R_{min}^2 (1 - \cos(\eta)), \qquad (7.8)$$

where $\eta = 180 - 2\omega$.

In triangle **1-2-3**,

$$h^2 = L^2 + \delta_r^2,$$

$$\sin(\mu) = \frac{\delta_r}{h}, \qquad (7.9)$$

$$\cos(\mu) = \frac{L}{h}.$$

Solving for R_{min} using the above set of equations, we get,

$$R_{min} = \frac{L^2 + \delta_r^2}{2\delta_r} \qquad (7.10)$$

According to this relation, for a given torchtip to sensor distance $(L > \delta_r)$, the R_{min} is inversely proportional to the operating range of the sensor. To ensure robust tracking, the radius of curvature of the seam R should be greater than R_{min} at all points along the seam.

7.3.3 Torch-Sensor Assembly Design and Seamwidth Considerations

An important concern while seam tracking is correctly positioning the scan window over the seam. It is desirable to have the scan lie completely across the seam in order to view the seam's characteristic features. Even when the scan width is greater than the seam width, there may be situations when a curving seam does not allow the entire seam section to be viewed. For a specific seam tracking system design, achieving this is a function of (i) the rotation of the sensor about the torch axis, and (ii) the seam's radius of curvature. The method to analyze the minimum allowable radius of curvature of the seam for proper scanning is discussed below. We assume that the seam lies in the xy-plane of the Torchtip frame and the scan center is positioned over the root of a seam with width *(2c)* (refer to Figure 7.5).

As shown in Figure 7.5, let the torchtip be positioned at the workpiece origin (0, 0). In the configuration shown, the sensor scans the seam (width $= 2c$) at a distance L ahead of the torchtip. The sensor's scan

166

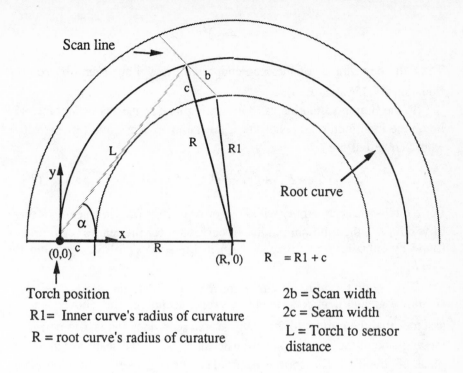

Scan line

b

c

R

R1

L

y

α

x

(0,0) c

R

(R, 0)   R = R1 + c

Root curve

Torch position

R1 = Inner curve's radius of curvature

R = root curve's radius of curature

2b = Scan width

2c = Seam width

L = Torch to sensor distance

Figure 7.5 Minimum radius of curvature based on the ratio of scanwidth to seamwidth.

(width = 2b) is centered on the root curve (radius = R, center (R,0)). Let the coordinates of the scan center be (x_b, y_b). From geometry, we get

$$x_b = \frac{L^2}{2R}, \qquad y_b = \frac{L}{2R}(4R^2 - L^2)^{1/2} \qquad (7.11)$$

Since the scan-line is normal to the torch-sensor vector **L**, the equation of the scan-line in normal form is given by Equation (7.12).

$$x \cos \alpha + y \sin \alpha = L \qquad (7.12)$$

where,

α is the angle made by the torch-to-sensor vector with the x-axis of the workpiece frame.

The endpoints of the scan-line (x_e, y_e) lie on the scanline thus satisfying Equation (712). On the other hand, the distance from each scan endpoint to the scan-center is the scan halfwidth(= b) and is given by Equation (7.13).

167

$$\left[\left(x_b - x_e \right)^2 + \left(y_b - y_e \right)^2 \right]^{1/2} = b \qquad (7.13)$$

Thus, the coordinates of the scan endpoints (x_e, y_e) are found by solving Equations (7.12) and (7.13).

In the limiting situation as shown in Figure 7.5, the scan endpoints will lie on the inner/outer curves of the seam (radius = $R \pm c$ and center $(R, 0)$), whose equations are

$$(x - R)^2 + (y)^2 = (R \pm c)^2 \qquad (7.14)$$

For a torch-sensor assembly with design parameters $(L, 2b)$ tracking a seam of width $(2c)$, the minimum radius of curvature for the seam's root curve is found by substituting $(x = x_e, y = y_e)$ in Equation (7.14) and solving for $R = R_{min}$.

The seam's radius of curvature R at all locations along the seam should be greater than R_{min} to ensure proper scanning. The above analysis assumes that the normal to the seam at the scan location is also normal to the xy-plane of the Torchtip coordinate frame, i.e., the seam lies in the xy-plane of the Torchtip coordinate frame. For any other orientation of the seam's normal, the projection of the seam's width in the xy-plane is smaller than the actual width $(2c)$ and hence, the above analysis represents a conservative approach.

Another issue related to viewing vee-groove seams is the relation between the vee-groove semiangle and the view vector orientation. For a range sensor with a fixed baseline, the limiting angle between the view vector and the vee-groove normal (in the xz-plane of the Seam coordinate frame) is a function of the vee-groove geometry. Although it is desirable to view the seam along the normal, as in Figure 7.6(a), this is not always possible. Excessive view vector angles may prevent proper range imaging of the seam because the incident/reflected beam may be occluded in some section of the scan, as in Figure 7.6(b). During range imaging, if the view vector angle (α) is less than the semi-vee angle (θ), then the incident beam can reach all points in the vee-groove. However, if this condition is not met then only one side of the vee-groove can be observed. This problem is most prevalent in a range sensor with a single camera. Some researchers (Waller et al. 1990) have circumvented the problem of occlusion within a vee-groove seam by using serial beam range sensor with two beam sources. The advantage of such a system lies in the fact that any point within the vee-

168

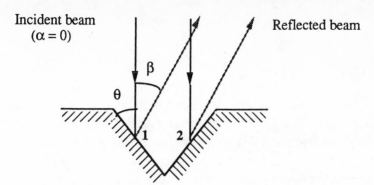

(a) Proper view-vector: Point 1 visible from both source and camera.

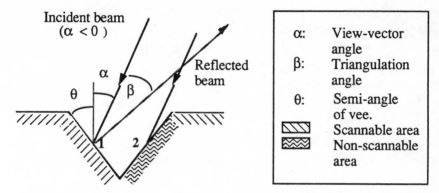

α:	View-vector angle
β:	Triangulation angle
θ:	Semi-angle of vee.
(hatched)	Scannable area
(wavy)	Non-scannable area

(b) Excess view-vector: Point 2 not visible from camera.

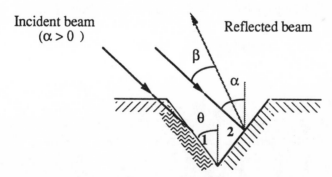

(c) Excess view-vector: Point 1 not visible from source.

Figure 7.6 Relation between the view vector orientation and the semiangle of a vee-groove.

169

groove can be illuminated by one of the two light sources and subsequently viewed by the single camera

7.4 SUMMARY

For robust seam tracking, it is important that any tracking error be bounded and not be allowed to diverge. It can be argued that by sensing the location of the torchtip mounted at the endeffector, and imaging the weld joint geometry during each cycle, the subsequent torch path generated is stable. The maximum tracking in any cycle is the upper bound of the tracking error for the entire tracking operation.

The seam tracking accuracy is affected by a variety of factors that can be basically categorized into those affecting the pathpoint computation and those affecting the actual pathpoint tracking. In the first category are factors such as (i) error in calibrating the end-of-arm tooling and other coordinate frames (e.g. torch and sensor frames), (ii) resolution of the sensors such as the robot arm joint encoders and the range sensors, (iii) inherent inaccuracy in the robot model, and (iv) errors in modeling the seam geometry. Errors in tracking the path are primarily due to (i) tracking nonlinear seams with piecewise linear segments, and (ii) robot inaccuracy in attaining the actual endeffector location corresponding to the commanded pathpoint in cartesian frame.

Reliable and accurate seam tracking also requires an understanding of the inherent limitation of any system's ability to track a seam, which is primarily a consequence of the system's design parameters. For tracking straight seams (i.e., radius of curvature tending to infinity), system requirements are relatively simple. As the seam's radius of curvature decreases, the requirements on the system performance increase and below a certain radius of curvature, the system can no longer track the seam within the allowable tracking error. The analysis presented here represents a preliminary investigation into the performance limitations of a seam tracking system. The limiting performance has been analyzed in terms of the system parameters such as (i) the maximum tracking speed in cartesian frame and the control-cycle period, (ii) the range sensor's field of view for a given torch-sensor assembly configuration, and (iii) the range sensor's scanwidth and the width of the seam cross-section.

In the preceding chapters, we have attempted to present a broad picture of the issues in intelligent seam tracking as well as the details of our solution approach and its results. The next chapter will acquaint the reader with other systems developed for adaptive, realtime, intelligent seam tracking for robotic welding. The reader may find it interesting to note the similarities and the differences in the innovative approaches used by several researchers/developers to address this problem through the use of various sensing techniques, seam tracking strategies, etc.

APPENDIX D

D.1 ROBOT ACCURACY ISSUES

Sensor-guided robots that receive offset information from sensors, and offline programmed robots that receive absolute position commands from a CAD system are limited by the robot accuracy. Robots need to be "accurate" in the sense that the physical location or trajectory reached by the endeffector is precisely where the robot has been directed to. Robot accuracy has been defined in a variety of ways. Robot accuracy as defined here is a measure of the difference between the actually attained position/path of the robot endeffector and the input position/path (in absolute world coordinate frame) commanded by the robot controller. By understanding the types of errors possible and the factors affecting the robot accuracy, one can devise methods to make compensations.

D.1.1 Factors Affecting Robot Accuracy

Robot accuracy can be influenced by a number of factors which are usually classified into the following categories:

1. Parametric factors;
2. Measurement factors;
3. Computational factors;
4. Application factors;
5. Environmental factors.

The contribution to the robot inaccuracy in each category may not be totally independent from one another. For instance, temperature change (an environmental factor) affects parametric factors (e... link length, friction coefficients) and also measurement factors (e.g. drifts in control electronics and measurement sensitivity).

D.1.1.1 Parametric Factors. Robot parametric factors include both kinematic parameters and dynamic parameters. Inaccurate representation of the robot parameters directly affect the robot accuracy. Generally, kinematic parameters affect the robot's *pose* accuracy while the dynamic parameters affect the *trajectory* accuracy. Kinematic parameters associated with the robot arm include link lengths, offset

172

distances between each pair of adjacent joint axes and angularity between each pair of adjacent axes. Link lengths and offset distances can vary from their nominal design specifications due to temperature variations, manufacturing errors (machining tolerances) and assembly errors (tolerance stackups). For a three-jointed robot, the endeffector position could vary by as much as ± 0.255 mm, assuming normal machining tolerances of ± 0.05 mm and tolerance stackups of RMS ± 0.085 mm (Day 1988). Assembly errors such as improper fitting and improper seating of bearings supporting the axes can cause angularity between each pair of adjacent joint axes. An error of just 0.1 degree in angularity could cause a link offset error of 1.745 mm when adjacent axes are 1000 mm apart. Similar link offset error can be generated due to incorrect zero-angle positions of the robot joints. This usually is a result of imprecise seating of the encoder/resolver or due to improper procedure for determining the zero angle position.

One of the most common sources of robot inaccuracy is the use of "ideal" robot design. Ideal robots have closed form inverse solutions such that desired robot endeffector motion in the cartesian frame can be converted to joint angle commands directly. This approach allows realtime computation in the robot controller to follow desired programmed paths in realtime. In reality, however, errors in the kinematic parameters due to environmental or manufacturing conditions often make the ideal robots "nonideal" thus, introducing robot inaccuracy. Nonuniformity in the motion of drive train components such as gears, chains and belts also introduces errors in the robot workspace. Methods suggested by researchers to correct this source of robot inaccuracy involve careful measurement of the error in gear train motion and subsequently mapping it into the robot controller software.

Dynamic parameters such as structural flexibility, inertial parameters of the robot, and frictional parameters in the drive train affect the robot control and accuracy. Structurally, robots are designed to be more flexible than machine tools and hence can deflect under their own weight. The various components contributing to the robot's flexible structure include the links, bearings and the drive train, especially when the payload at the endeffector changes. The links usually have a box-section or I-section to maximize the sectional moduli of elasticity and minimize the deflection. Bearings are designed to not only carry loads at specified speeds but also to have small angular deflection under moment loading. Drive train components are designed to provide overall stiffness within

173

reasonable value. However, each one of the above factors contributes to the overall flexibility of the robot structure and hence to the robot inaccuracy.

Inertial parameters of the robot play a role in the trajectory accuracy and velocity accuracy of robots. Inertial parameters are not precisely measurable in robot manipulators and can change due to factors such as variation in the casting thickness and thermal distortion during welding. It is estimated that inertial parameters can vary by 10 percent or more from robot to robot of the same model (Day 1988), making robot control quite challenging.

Friction parameters of robots are difficult to quantify accurately. All friction parameters, such as viscous friction, stiction, and coulomb friction dissipate energy and contribute to the robot inaccuracy. In most nondirect-drive geartrains, between 10–50 percent of the motorshaft output power is lost to friction before it is delivered to the finally driven joint.

D.1.1.2 Measurement Factors. Encoders and resolvers are the most common position feedback devices used for robot control. The resolution of these devices can range from several hundred counts to several hundred thousand counts per revolution. These position feedback devices can be mounted either on the motor shaft or on the joint axes. Although mounting the encoders on the motor shaft provides ease of motor commutation control, the absence of feedback about flexibility, backlash and other parametric factors affects robot accuracy and so is a serious disadvantage. On the other hand, encoders mounted directly on the joint axes provide good feedback to the robot controller, but require high resolution to detect small changes in joint positions.

Compliance of the drive train between the motor and the rotated joint can affect robot performance and accuracy especially at high loads. Robot flexibility can be accurately controlled by sensing deflection of the links through strain gage sensors mounted at appropriate places on robot links. Such methods can help in reducing endeffector inaccuracy. Alternately endeffector inaccuracy can be controlled using load sensing to compute flexibility compensation. It should be noted, however, that all these techniques impose strict resolution requirements on the sensors. The measurement resolution of the sensors should be better than the accuracy required of the application.

The measurements from internal sensors such as encoders and resolvers are affected by the parametric factors of the robot. An alternative approach is to measure the endeffector position independent of the robot parameters. This involves using external measurement instruments such as laser interferometers, theodolites, coordinate measuring machines, etc. Some of the external measurement devices provide static measurements only and are often used for calibrating the robot's pose characteristics. Other devices with high data sampling rates can provide feedback to the robot controller for accurate position tracking in realtime. Example of such a system is the one developed by Lau and coworkers at the National Institute of Standards and Technology (NIST). These measurement devices provide more reliable and accurate data than the robot's internal sensors because their resolution requirements are usually quite high. Tracking laser interferometer can provide accuracies of 0.1 mm while static laser interferometers and coordinate measuring machine are accurate to 0.02 mm.

D.1.1.3 Computational Factors. The endeffector path as computed by the robot controller can contain various errors and thus affect the robot accuracy. These include roundoff errors, error in the kinematic model of the robot arm, and errors in computing the robot positions at or near the robot singularity positions.

Roundoff errors are generated since the robot controller uses a finite number of bits to represent numbers. For most computations, 16 bits will be enough to retain the desired precision in robot path computations. However, for those computations that accumulate small increments of motions, inaccuracy can be introduced due to roundoff errors.

As mentioned earlier, most robot controllers assume ideal robot kinematics since they allow the use of direct formulations for computing robot inverse solution. However, in reality the manipulator arm is seldom built with ideal kinematics although the design may be so. This error is in the range of manufacturing tolerance, usually ± 0.1 mm. The other possibility is to use nonideal formulations for computations involving nonideal robot arms. However, nonideal formulations require much more computation and would prevent realtime control of the arm through today's robot controllers. The robot accuracy is also compromised around singularity positions in order to maintain stable robot arm motion.

D.1.1.4 Application Factors. Besides the factors inherent in the robot itself, the accuracy desired within an application can be adversely affected due to a variety of reasons. These include:

(i)   *Installation Errors,* such as installation on soft ground, etc. can mean inadequate dimensional control of the robot installation. This will introduce errors in the world coordinate frame of the robot and consequently affect all the other coordinate frames.

(ii)   *Workpiece dimensional integrity.* For those applications requiring the workpiece dimensions to be within strict tolerances, tolerance stackups and other manufacturing errors of the workpiece can place undue performance requirements on the robot.

(iii)   *End-of-arm tooling.* Errors in the design, manufacturing, and installation of the end of arm tooling can adversely affect the robot accuracy. In the case of seam tracking systems, error in the definition of the torchtip coordinate frame can result in error while positioning and/or orienting the torch along the seam. Attention should be paid to this factor which in many cases is cheaper to correct than robot inaccuracy.

D.1.1.5 Environmental Factors. Temperature is perhaps the most important environmental factor affecting robot accuracy. Temperature variations affect the link lengths in the structural and drive train components of the robot arm. For example, a 10 F change in the ambient temperature can cause a 600 mm steel ball screw to change by 0.048 mm. If the length of the endeffector is 8 times that of the moment arm of the ball screw, actual change of endeffector accuracy can be (0.048 x 8 = 0.384 mm) (Day 1988).

Temperature also affects the characteristics of lubricants used in robots. It can also cause the electrical parameters of such analog components such as resistors, capacitors, and operational amplifiers to drift. With robot joint encoders, an offset in their nominal values can cause a change of upto 0.01 mm at the endeffector.

D.1.2 Methods to Improve Robot Accuracy

The methods for improving the robot accuracy can be classified into the following three categories: Robot calibration; Open loop compensation; and Closed loop position feedback.

176

D.1.2.1 Robot Calibration. Industrial robots move to programmed positions as directed by the robot's control equations. Control equations such as inverse kinematics equations for robot motion control are derived from the mathematical model of the robot joint and link structure. For accurate robot moves in the cartesian space, the zero reference position of the robot joints corresponding to a predetermined robot pose should be precisely known. Since the robot controller receives feedback about joint angles through encoder values, the relationship between encoder values and the zero reference position of the joint angles should be determined for the predetermined robot pose. The process of establishing these relationships and reference positions is known as Robot Calibration. Calibration procedures help establish accuracy at one particular position in the robot workspace to the order of ± 0.025 mm. Hence, it is one of the first order correction to improve robot accuracy.

Robot calibration can be accomplished through several methods. The predetermined robot pose can be established by using external sensors, calibration fixtures, or by physical markings on the robot structure. Using physical markings on the robot structure, however, may not be consistent from robot to robot. Hence, external sensors or calibration fixtures (known as mastering fixtures) are preferred. Once the pose is attained, the encoder values recorded establish the common reference between the encoder values and the robot joints.

The need for users to go through the calibration process is eliminated by using absolute encoders instead of incremental encoders. This way, the manufacturer is responsible for determining the initial zero reference position.

In order to perform the intended task accurately, most robotic applications require the users to determine the workpiece coordinate frame, and in the case of sensor-guided robots, the sensor coordinate frame. The workpiece coordinate frame can be determined accurately, by using the external sensors in the world coordinate frame. Relative accuracy between the robot and the workpiece can be established by commanding the robot to move to three calibrated positions on the workpiece. A (4 x 4) transformation is then calculated to transform from the robot coordinate frame to the workpiece coordinate frame or vice versa. It should be noted that the transformation still contains the uncorrected robot parameters. The process of establishing the sensor coordinate frame relative to the tool coordinate frame depends on the application.

177

Methods to define the sensor coordinate frame for seam tracking applications have been discussed in Chapter 4.

D.1.2.2 Open Loop Compensation. To improve the robot endeffector accuracy, the open loop methods make compensated commands to the robot based on some predetermined information about the robot and factors contributing to the inaccuracy. Unlike the robot calibration process which improves accuracy at only one particular position in the robot workspace, the open loop compensation is intended to improve accuracy in most of the robot workspace.

Open loop compensation comprises of three tasks: measurement of robot accuracy, error modeling, and error compensation. Measurement and the error modeling tasks are performed offline while the error compensation is done online within the robot controller while commanding robot moves.

The measurement of robot accuracy allows one to understand what errors are to be corrected. Two sets of measurements are made at various locations in the robot workspace. First the robot endeffector positions are measured by an external sensor such as a laser interferometer or a theodolite system in the world frame. The other set contains the corresponding robot positions as commanded by the robot controller. Proper transformation between the two robot positions yields the actual error in robot accuracy. Unfortunately, this task is quite tedious and may require several man-months of effort in data collection and verification. Unpublished robot inaccuracies measured by users and manufacturers are in the order of 5 mm to 10 mm for typical robots (Day 1988).

The next step in the open loop compensation method is error modeling in order to determine the actual robot arm structural parameters. Models ranging from pure kinematic parameter correction to full models containing drive train and backlash corrections have been proposed by researchers [Davies, et al. 1990]. Many of these models assume small parameter errors in the global robot model in order to allow the correction parameters to be linear in the equations. The corrected parameters are calculated using least square fit or through minimization of some global error function. An alternate approach proposed by Davies and coworkers [Davies, et al. 1990], attempts to improve the local accuracy of the robot at the distal link measured relative to the workpiece in localized workpiece regions. By applying the relative calibration methodology to a

178

GMF 2-200 six-axis robot, the authors report a reduction in the average localized robot inaccuracy from 3.4 mm to 0.86 mm.

In the final task of error compensation, the compensation is made by the robot controller by commanding moves based on the corrected parameters. The compensated commands are easy to compute while working with ideal robot configurations. Nonideal configurations containing parameters such as offsets may require the computation of robot Jacobians as a part of the path planning itself. Considering the computational complexity involved, most present day robot controllers will find this impractical.

Open loop compensation using corrected robot parameters can improve the robot accuracy 5 to 10 times or about 1 mm in the robot workspace. Theoretically, one could get better results by incorporating more parameters in the error model, such as load deflection model, etc. However, there is a limit to the number of parameters the robot controller can handle. Small robots with their limited workspace can be made more accurate but accuracy better than 0.5 mm is still difficult to achieve. Although open loop compensation based on measuring static positions of the robot endeffector can correct for kinematic parameters well, the dynamic parameters cannot be fully corrected. Dynamic parameter identification and control is still very much a research issue.

D.1.2.3 Closed Loop Methods. The open loop compensation provides corrected robot parameters based on a "best fit" approach to the data at various measurement locations. The robot moves are commanded using these corrected robot parameters. Still, there is error left at the measured points and likely more so at other locations in the workspace. The open loop method does not have any additional feedback of robot accuracy. Additionally, there is a limit to how many parametric compensations the robot controller can actually handle. Hence, in order to improve the position control of the endeffector even further, the closed loop approach monitors the robot endeffector position and orientation in the world frame using external sensors. The data sampling rates of the external sensors are high enough to integrate their measurements for closed loop control of the robot endeffector position. This approach allows the robot accuracy to be improved over the entire robot workspace. The laser tracking system developed by Lau and coworkers is an example of the closed-loop the endeffector position control that allows the accuracy to be improved to 0.012 mm within a 2 m workspace. Today the accuracy achieved based on the closed-loop method by integrating the laser tracking system with the robot controller is about the best accuracy one can achieve. The reader is referred to (Lau et al.1985) for details on robot motion tracking systems.

CHAPTER 8

SEAM TRACKING SYSTEM PROTOTYPES

8.1 ULTRASONIC RANGING-BASED SEAM TRACKING

Overview. Umeagukwu and coworkers at the Georgia Institute of Technology, USA report the development of a robotic seam tracking system based on ultrasonic range sensors (Umeagukwu, et al. 1989). The initial prototype is designed to track 2D-seams in a nonwelding environment. Two seam types tested include lap joints and vee-joints.

The primary intention of the authors in using ultrasonic ranging methods is to capitalize on its advantages over vision and through-the-arc methods. Although vision-sensing gathers data independent of the welding system, it requires complex signal processing to extract and interpret seam features from a noisy range image. Additionally, vision sensors react adversely to material conditions such as presence of rust, shiny polished surfaces, etc. Through-the-arc sensing also captures arc noise in the form of signal fluctuations. Other limitations, such as the purposeful weaving of a through-the-arc system, make it difficult to track sharp curves. Although ultrasonic sensors are also affected by the arc noise, their attraction lies in the ability to glean range information without the computational complexity of vision sensing. Researchers have demonstrated that ultrasonic sensory techniques are feasible for seam tracking and quality control. (Estochen, Neuman, and Prinz 1984; Fenn and Stroud 1986).

System Architecture. The architecture of this seam tracking system designed by Umeagukwu and coworkers is shown in Figure 8.1. It uses a Krautkramer 3-20 air coupled transducer, a Krautkramer LAM 80/8 sound distance meter for sensing the seam, and the positioning mechanism is installed on a GE P-50 process robot.

Ultrasonic Ranging Subsystem. The measurements are made by insonifying the workpiece surface at 45 degrees at a frequency of 100 kHz

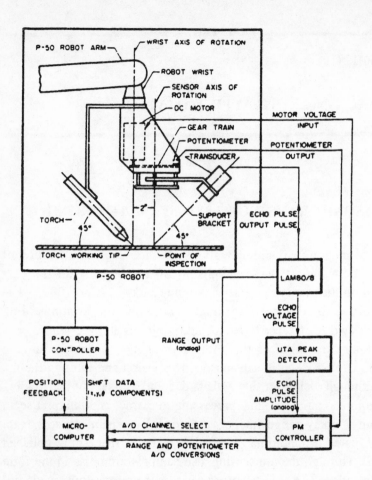

Figure 8.1 Architecture of the ultrasonic ranging based seam tracking system.[†]

with a wavelength of 0.33 cm in air. The maximum amplitude of the echo identifies that the sensor axis is aligned normal to the seam surface. Seam tracking requires computation of the seam's position and orientation in relation to the torch. The transmit time determines the location of the seam in relation to the sensor. The seam tracker measures both the time-of-flight and amplitude of the ultrasonic echo using standard electronic circuits resulting in a simple seam tracking algorithm. However, while

---

[†] Reprinted from Umeagukwu et al., Robotic acoustic seam tracking–system development and application, IEEE Trans. on Industrial Electronics Vol. 36 No.3, 1989, © IEEE 1989.

measuring the seam at 45 degrees, the variations in the pulse amplitude are symmetrical about the normal to the object. This makes it difficult to determine the side of the normal on which the sensor is located. The authors report to have overcome this problem by purposely oscillating the sensor about its axis of rotation at a rate of 2.1 Hz and an amplitude of 5 degrees.

The mechanism supporting the torch-sensor assembly is mounted on the endeffector such that both the torch and the sensor are simultaneously subjected to any wrist rotation. In addition, the sensor mechanism can independently rotate the sensor about a vertical axis. The oscillations of the sensor about the vertical axis demands that the same point on the workpiece surface be insonified throughout. This can be achieved on a 2D planar seam by ensuring that the vertical axis and the sensor axis intersect at the workpiece surface. Unfortunately, for 3D seam tracking in an unstructured environment, this assumption will normally not hold true.

At any instant, the sensor orientation in relation to the torch is measured using a potentiometer mounted on the mechanism. During the sensor oscillation, the pulse amplitude increases and reaches the maximum peak when the sensor axis is aligned normal to the seam surface. Therefore, the pulse amplitude provides the seam orientation in relation to the transducer. Likewise, the range measurement provides the distance to the seam which is converted into joint deviation based on a joint type specific formula and other joint parameters.

Seam Tracking. Tracking is performed by bringing the torch to its initial position of a taught location (P_0). Seam search is then initiated to compensate for any workpiece positioning error. The seam between the torch initial position and the lookahead sensor is assumed to be straight, and (n-1) shifts are calculated to position the torch at the intermediate locations. The data sampling begins with point P_n. Once P_n is sampled, data pertaining to shift dP_1 (vector joining points P_0 and P_1) is transferred to the robot controller. While the shift is being executed, the shift dP_n is computed before torch reaches point P_1. By keeping the computation of dP_n within a shift cycle, the shift dP_1 can be transferred to the robot controller as soon as the point P_1 is reached.

The authors report to have successfully tracked a planar vee-groove milled on an aluminum plate with depths of 0.318 cm. and sides cut at 45 degrees. The overall accuracy of the range sensor is reported to be ± 0.51 mm and orientation measurement accuracy to be within ± 0.5 degrees.

However, results on the overall seam tracking accuracy are not reported in the literature.

Summary. Seam tracking based on ultrasonic range sensors, although seemingly advantageous, has shortcomings especially in welding environments. Although the basic ultrasonic ranging is quite simple, its application to seam tracking is apparently complex. Measuring the joint orientation requires that the normal to the seam surface be identified. However, this can be difficult using maximum pulse amplitude if the sensor is not positioned correctly along the seam. The pulse amplitude itself can be affected by the echoes from areas neighboring the seam. Further, the sensor oscillation for detecting the maximum pulse amplitude suffers from symmetricity around the normal, and if the vertical axis and sensor axis do not intersect at the point being insonified on the workpiece then the oscillation may sweep across several points instead of a single point.

The use of ultrasonic-based ranging in arc welding environments brings up some unique concerns. For example, the thermal gradient in the air during welding and the presence of shielding gas reduce the speed of sound thereby giving incorrect range values. The use of a flexible barrier to shield the sensor provides only a partial solution at best, since the sensor is still affected by the shielding gas which reduces the speed of sound due to its large molecular weight (Umeagukwu et al. 1989). As a result, the range value has an offset, which can possibly be maintained fairly constant with a good seal between the workpiece and the skirt. Many researchers, however, have rejected the use of a skirt in arc welding environments. Using a horn concentrator positioned with extreme precision to reduce the variations in the return signal is a possible solution. However, the authors report that they still encountered some variability at times and further work is required to qualify ultrasonic range sensors within an arc welding environment.

8.2 AUTOMATIX INC.'S VISION-AIDED ROBOTIC WELDER

The vision-aided robotic welder developed by Automatix Inc., Massachussetts, USA, is based on visually determining 3D offsets from a taught path and incorporating them in realtime for path correction (Agapakis, et al. 1990b). In addition, it supports such features as offline detection of large fixturing errors before the start of welding (part finding) and inprocess adjustment of welding process parameters on the basis of monitored cross-sectional dimensions (adaptive welding).

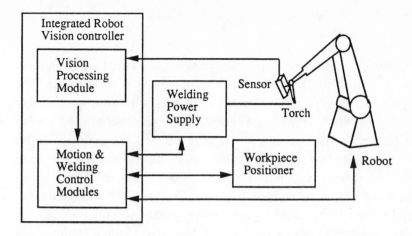

Figure 8.2 Architecture of the vision-aided robotic welder by Automatix Inc.[†]

The Automatix robotic welder is unique in its use of structured light in the form of a steerable cone of laser light. The implementation features multiprocessing computer hardware and software, wherein different processors share the necessary realtime computations for robot motion trajectory planning, multi-axis servo control, 3D workpiece surface modeling, and robot plus welding process control based on sensory feedback.

System Architecture. The system architecture of this robotic welder is illustrated in Figure 8.2. The preview sensor views the weld joint ahead of the welding torch as the robot travels along the seam. The vision processing system analyzes the sensory images to determine the location of the weld seam in relation to the sensor and also the cross-sectional dimensions of the weld joint. Based on this information, the motion control module corrects the welding torch path while the weld process control module adjusts the welding process parameters to compensate for part or joint geometry variations. User access and interfaces are provided at multiple levels, including, a handheld teach pendant; menu-driven

---

[†] Reprinted from Int. Journal of Robotics Research, Vol. 9:5, 1990, Agapakis et al., Vision-aided robotic welding: an approach and a flexible implementation, by permission of the MIT Press, Cambridge, Massachusetts, Copyright Massachusetts Institute of Technology 1990.

interfaces; RAIL–robot and vision programming language; and other application support libraries.

Preview Sensor. This subsystem consists of a CCD image sensor and an optical projection system that generates a section of conic light surface. The source of this monochromatic structured light is a 10 milliwatt He-Ne laser. A narrow band, optical interference filter centered around He-Ne laser frequency is used to block out ambient and welding arc light. According to the authors, the conic light surface has certain advantages in the context of robotic welding. The conic light surface requires a radial image scanning approach that is particularly effective in discriminating real laser stripes from flying welding arc spatter emanating radially from the torch. The sensor is hardened to operate in the harsh welding environment by packaging it in a compact, dustfree, constant temperature, and positive air pressure environment. The torch and the sensor are attached to the robot endeffector using an omnidirectional, resettable release bracket employing rotational frictional joints. In the event of an accidental collision, the friction joints yield without hysteresis and the motor power is switched off to prevent further damage.

Vision Processing. The vision processing module supports stripe detection, stripe contour analysis and feature recognition, modeling 3D geometry of the weld surface, and image-to-world transformation of the detected feature points. Stripe detection starts with intensity thresholding wherein the threshold is dynamically computed along the scan line based on cumulative intensity histogram. In the stripe contour shape analysis and feature recognition step, feature points and feature surfaces corresponding to weldjoint features are identified in the laser stripe image. Such stripe features can be recognized as long as the general expected shape of the joint can be described as a collection of generic feature points and surfaces (bottom-up approach to seam recognition). Therefore, the description of all standard welding joints (fillet, square-butt, left and right lap, vee-groove, etc.) are built into the system. In the final stage of vision processing, the 3D geometry of the weld joint surface is described on the basis of information from successively recognized laser stripes. This model provides detailed measurement on the seam's cross-section and also a description of the seam's coordinate frame.

186

Robot Motion Control. The robot motion control software can support single arm and multiple arm configurations. The hierarchical control architecture of this system has multiple layers supporting seam tracking control tasks, device motion planning tasks, and joint servo control tasks. The device's motion planning task plans the trajectory for the robot and derives the interpolation points 80–100 ms apart along each preprogrammed path segment. The root position computed by the vision processing algorithm is compared with the taught path to find out where along the path it is located and how far it is from the taught path. Thus, offsets are computed from the taught path at discrete interpolation points in each path segment. The joint space angles, computed using inverse kinematic solution of the arm, are sent to the servo task which controls the position of each axis. The servo task interpolates position points in joint space at a rate, typically, around 100 Hz.

The seam tracking control tasks are executed synchronously with the robot motion control tasks. For every new commanded position of the robot at the interpolation point, a vision offset, arm speed, a set of weld parameters are generated and sent to the robot and welding equipment.

Offline Part Finding. Seam tracking is done in conjunction with part finding, which entails using the robot mounted sensor to search for the part prior to welding. This step allows identifying gross part fixturing errors or dimensional errors that could prevent the seam tracking system from finding the start of the seam. The error can be modeled as a rigid body transformation between a model part and the actual part, and be used to uniformly offset all preprogrammed robot paths on the part.

Adaptive Weld Process Control. Welding process parameters such as arc voltage, wirefeed rate, torch travel speed, and torch position relative to the seam are adjusted in realtime based on previewing weld joint features and cross-sectional dimensions ahead of the welding arc. Proper weld process control requires understanding the significance of the variations of weld joint features and cross-sectional dimensions on the welding process and breaking them down into subranges. This knowledge can be stored in the welding adaptation table and used to lookup combinations of welding process parameters appropriate for each possible combination of cross-sectional measurements.

187

8.3 PASS – VISUAL SENSING FOR TRACKING AND ONLINE CONTROL

Overview. R. Niepold and F. Brummer at the Fraunhofer Institute, West Germany report the development of a visual sensing system for seam tracking and online process parameter control in arc welding applications (Niepold and Brummer 1987). The sensing system is called PASS which stands for Programmable Adaptive Sensor System. PASS processes standard video signals from TV cameras with resolution up to 512 x 256 pixels, and the prime task is to identify geometric characteristics (eg., location, shape, and dimension) of the weld pieces. The basic version of PASS has been in operation since 1980.

System Architecture. PASS is capable of accommodating up to four CCD TV cameras and their signals can be processed in parallel (refer to Figure 8.3). Binary images are derived from the video signals by use of adaptive thresholds. Picture elements are selected by the model hardware processors, which are programmed using a compressed model of the expected scene. The selected picture elements are analyzed by the

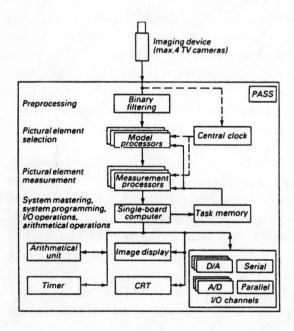

Figure 8.3 Functional block diagram of PASS. (From Lane 1987)

188

measurement hardware processors. Their function is determined by programming and can be altered or repeated during image processing. The system automatically adapts the involved models depending on the scene evaluations.

PASS is supplied with a modular software library where the individual subprograms represent specific functions such as image processing and control modules. All the data needed for a specific task may be stored in a permanent task memory, from which they can be easily called. Specific operating parameters are programmed using a menu technique.

Basic Features. The major features of PASS are as delineated below:

Speed of Operation: The processing period is scheduled not to exceed 40 ms (i.e., the sampling frequency is no less than 25 Hz) because the sensing system may operate as an element of the feedback control system. Within this time interval, a singleshot image has to be processed down to the required measurement values, including control computations and input/output operations.

Flexibility: Since variations in the structure of the monitored scenes could be significant for different applications, fixed image interpolation strategies are not suitable. The programmable capability of PASS offers flexibility to deal with the individual applications.

During welding, PASS records the signals in its memory. An identification algorithm that belongs to the PASS software library is executed offline to obtain a mathematical description of the process to be controlled. PASS provides torch positioning and process control based on monitoring of the weld spot. For example, when welding butt seam profiles, the sensor system measures the gap width which is reflected by the characteristic excrescence in the monitored melting pool. The algorithm consists of two major operational steps:

1. *Detection of the Seam Line*: A linear regression algorithm processes the data of detected points to generate the lateral position and orientation of the seam line within the camera coordinates. The distance between the torch (supporting the imaging device) and the workpiece surface is obtained by correlating the monitored pattern with a reference pattern stored previously along the vertical axis of the field of view.
2. *Measurement of Profile Geometry:* With butt seam profiles, the gap width is obtained by computing the mean distance between the

189

detected points and the previously determined seam line. Overlapping seams may show a gap between the upper and lower sheets. The gap width is calculated by correlating the two main patterns along the vertical axis of the frame of view taking the strength of the upper sheet into account. In general, PASS serves as the sensor system of the whole welding installation.

8.4 ASEA ROBOTICS' ADAPTIVE TORCH POSITIONER

Overview. T. Porsander and T. Sthen of ASEA Robotics AB, Sweden report the development of an adaptive torch positioner and seam finder that permits robot-assisted welding of workpieces (Porsander and Sthen, 1987). This sensing system is intended mainly for sheet metal welding where weld lengths are short and starting points are not difficult to search. Besides searching the joint, the system also measures the gap at overlap joints thereby allowing the robot to automatically adapt the welding parameters to the actual gap.

System Architecture. The major components of the adaptive torch positioning and seam finding are shown in Figure 8.4 and include:
- An optical sensor with control electronics;
- An IRB 6AW or IRB L6AW robot equipped with a control system for executing adaptive functions;
- An integrated microcomputer system which evaluates the sensor signals and transmits the validated data to the adaptive functions in the robot control system.

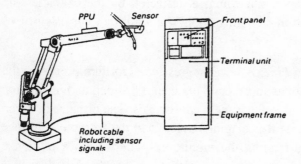

Figure 8.4 ASEA's adaptive positioner. (From Lane 1987)

190

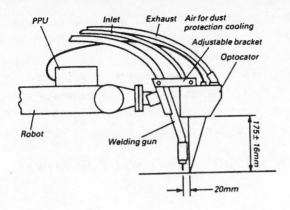

Figure 8.5 Sensor head arrangement of the ASEA adaptive positioner.
(From Lane 1987)

As illustrated in Figure 8.5, the sensor is compact, environmentally protected, and mounted on the torch with the measuring spot approximately 20 mm from the wire tip. The standoff distance of the sensor head is 175 mm and has a range of 32 mm with a resolution of 0.06 mm. The control electronics for the sensor head are fitted on the robot arm. The robot controller is equipped with a plug-in sensor interface board and a special evaluation computer board.

Operational Features. When searching, the robot moves the sensor towards a surface or across an edge or a joint. The signal, i.e., the distance to the surface, is evaluated by the computer. When a certain distance is reached or an edge/joint is detected, a search stop signal is sent to the robot. The operations are based on the following criteria:

- Location of edges;
- Location of overlap joints;
- Location of fillet joints;
- Location of the elevation of a surface;
- Measurements of gap width in overlap joints for automatic selection of welding data.

In the searching process, the joint is defined in three dimensions and the welding gun is positioned simultaneously. A complete search in three dimensions and location of the gun takes typically less than 1.5 second. A

191

search in two dimensions is often sufficient. The search is performed without actuation of the arc. The robot is programmed with information relating to the material thickness and the search type (e.g., butt joint, overlap joint, or fillet joint). The welding parameters are autonomously updated. With overlap joints the system can detect sheet thicknesses down to 0.8 mm. The searching accuracy of the system depends on the search speed and is typically better than 0.4 mm.

8.5 A REALTIME OPTICAL PROFILE SENSOR FOR ROBOT WELDING

Overview. G.L. Oomen and W.J.P.A. Verbeek at Oldelft, The Netherlands, have developed a compact 3D vision sensor for robotic arc welding (Oomen and Verbeek, 1987). This optical sensing device generates data to correct the preprogrammed welding path, and has the capabilities of executing the welding operations. Mounted on the robot hand, the laser-based scanning triangulation sensor takes distance measurements to the workpiece at a rate of 2000 per second.

System Architecture. The major components of the optical sensor are a camera, a preprocessing unit and a microcomputer. The camera is water-cooled, weighs about 1 kg, and is protected against the effects of weld splatter and smoke. It contains a short He-Ne laser, deflecting mirrors, motor, receiving optics, detector array and relay electronics. It is connected by cable to the preprocessing unit which may be mounted on the robot arm. This unit contains the laser power supply unit, the reflection point detector, and a microprocessor. The processor controls the motor and performs linearization, transformation and data reduction to allow accurate transmission of the measured seam profile to the robot control unit over a cable up to 50 m.

Operational Features. The operating principle of the optical sensor builds upon the concept of triangulation where a laser beam is directed onto the object and the diffusely scattered light is sensed by an oblique detection system. The lateral resolution of the sensor can be made dependent only on the laser beam width which, in turn, can be optimized over the desired range by suitable optical means. Basic operational data of the sensor are listed below.

192

Sensor Technical Data:

Laser beam width	632 nm
External laser power	1 mW
Precision, lateral	0.2 mm
Precision, radial distance	0.2 mm
Maximum field of cross-sectional view	60 x 80 mm
Scan efficiency	80%
Number of depth measurements per second	2500 max
Number of scans (profiles) per second	10

The basic operational capabilities of the optical sensor are as follows:

- Finding the start of the weld seam;
- Calculating the seam volume for process control;
- Detecting tack welds;
- Locating the surface elevation ;
- Measuring the gap width in overlap joints for automatic selection of welding data.

Seam Tracking and Process Control Experiences. The sensor system has been tested to function properly under environment of splatter, smoke, deposits, stray light from welding arc, and electromagnetic interference. Since September 1984 this sensor system has been used with a Cyro robot in a large US aluminium construction plant. The tracking method was expanded for realtime coordinated table motion. The sensor system has also been used for sheet metal welding in automotive industry and also for observation of welded seams in the high-frequency butt welding process (1 MW, 200 kHz) of large diameter gas pipes. In general, the sensor system can be used for the following realtime applications:

- Generating data to correct the preprogrammed welding path;
- Finding the starting point of the seam;
- Calculating the seam volume for process control;
- Detecting tack welds;

8.6 LASERTRAK – SEAM TRACKER FOR ARC WELDING APPLICATIONS

Overview. R. Bjorkelund at *ASEA* Robotics, Sweden reports the development of a one-pass seam tracker, called LaserTrak, for arc welding applications (Bjorkelund, 1987). The LaserTrak is an optical noncontact sensor that is capable of tracking while welding different types of seams.

Operating Principle and System Architecture. The sensing element of LaserTrak is positioned around a standard welding gun, and functions on the principle of optical triangulation where a beam of light is projected onto the welding surface to reflect diffused light onto a position-sensitive opto-electronic device (refer to Figure 8.6). The position of the reflected beam on the device indicates the distance from the surface. Two mirrors, mounted on rotatable axes serve to generate the transmitted and reflected light beams. By rotating the axes, the light beams are moved to and fro ±15mm across the joint. The position of the seam is obtained by electronically analyzing the measured height profile. A scanning frequency of about 20 Hz is required to be able to accurately track rapid changes in the path of up to 45 degrees. At this frequency, the seam position can be updated every 0.5 mm. The scanning movement is regulated by changing the axes of the two mirrors using an electronically controlled

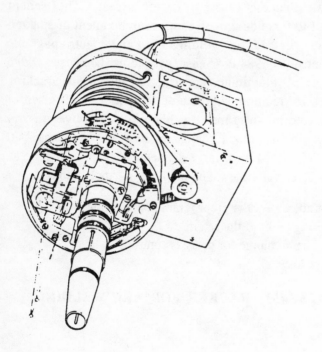

Figure 8.6 Bottom view of ASEA's LaserTrak system. (From Lane 1987)

194

electromagnet. To be able to measure ahead of the welding gun in whatever direction it moves, the sensor is fixed on a rotating disk that can rotate the sensor through 540 degrees.

In LaserTrak, the optical signal for height measurement is separated from the arc light in the following way. The optical energy source is a continuous-wave infra-red laser with a wavelength of 820 nm and an output power of 20 mW. A tuned optical interference filter that is transparent for a narrow range around the wavelength of the transmitted infrared laser beam is positioned in front of the optical receiver. The laser beam is modulated at 32 kHz and the receiving circuits are sensitive to this signal only. The laser beam is focused onto the surface such that the spot in vertical measuring range is within 0.3 mm in diameter, and the resolution of vertical measurements is 0.02 mm. The receiving device is a position-dependent diode unlike the commonly used CCD array.

The LaserTrak software is user-friendly. For example, a program for welding a seam with a starting point within a bound of ± 12 mm from a nominal position and with no corners greater than 45 ° requires only two programming instructions.

Basic Features. The major features of LaserTrak are as follows:

- Automatic vertical and horizontal correction of the path (including the errors resulting from thermal deformation).
- Flexibility of operations as a result of relatively less constraints on welding objects and their fixtures.
- Adaptability to welding parameter changes and variations.
- Ease of operations in terms of reduced programming time.
- Cost effectiveness in terms of low rejection rate of the welded products.
- Product enhancement in terms of high welding quality.

8.7 ESPRIT – REALTIME IMAGING FOR ARC WELDING

Overview. D.N. Waller and C.J. Foster of The Welding Institute, Cambridge, UK, and R. Wagner of Rhein.-Westf. Technical Institute, Reutershagweg, Germany, report the development of control schemes for automated welding machines under the ESPRIT (CIM) Project (Waller, Foster and Wagner, 1990). Two different realizations of intelligent welding system concepts have been studied: (i) Single-pass welding representative of the thin-sheet/high-volume industry (eg., automotive); and (ii) multi-pass welding, representative of the heavy-weight/low-volume industry (eg., power generation). The single-pass system provides user programmable process control while the control algorithms of the multi-pass system are embedded within the system software.

Basic Features of the Control Systems. The following broad classes of closed-loop control systems have been considered in the ESPRIT (CIM) Project:

- *System tracking for positional adaptability*: Systems having the ability to determine the course of the joint preparation and to produce a signal which can be used to correct the position of the welding head.
- *Flexibility* of operations as a result of relatively less constraints on welding objects and their fixtures.
- *Joint recognition for process adaptability*: System involving measurements of significant elements of the joint preparation to allow correction of the welding parameters.
- *Single-pass and multi-pass welding:* To ensure total system adaptivity (i.e., consisting of positional and process manipulation adaptability), welding tasks are divided into two groups, namely, single-pass and multi-pass welding. In single-pass welding, high productivity is attained by maximizing welding speed while adaptively controlling the welding process to compensate for variations in the joint geometry. In multi-pass welding, the joint is made up of sequentially deposited weld passes. Joint completion rate is dependent on the metal deposition rate allowed by the energy input limits.

8.8 SUMMARY

As is evident from the previous discussion, the majority of the seam tracking systems utilize vision sensing techniques for both, positioning the torch along the seam as well as for welding process control. Each seam tracking system distinguishes itself from the others through a variety of factors such as, capability to handle different joint types, frequency of range measurements, ease of use and programmability, adaptive welding process control capability, etc. In the following and final chapter, we discuss certain areas which are important to the future development of seam tracking and robotic welding technology.

CHAPTER 9

FUTURE DIRECTIONS

9.1 INTRODUCTION

The work discussed in this monograph represents an important aspect of the adaptive, intelligent seam tracking problem. Although the control scheme for tracking an unknown seam in an unstructured environment is fairly well-developed, the system performance can be enhanced in the following areas associated with tracking complex seams and adaptive welding.

- To further enhance the performance and the capabilities of the seam tracking system for welding automation, one could (i) increase the number of degrees-of-freedom of the system, and (ii) introduce an online welding planner.
- Additionally, to obtain a realistic estimate of the upper bound on the tracking error, a detailed analysis of the interaction among the various factors contributing to the tracking error is required.

The following sections discuss the possible approaches for future study in the above-mentioned areas.

9.2 SCOPE OF FUTURE ENHANCEMENTS

Enhancements to the system can be made at two levels: (i) improving the seam tracking capabilities of the system by introducing more degrees of freedom in the manipulating devices, and (ii) addressing issues related to welding automation such as integrating an online welding planner with the high-level controller.

9.2.1 System with Additional Degrees of Freedom

The seam tracking system can be made more flexible by (i) integrating a part positioner, and (ii) by increasing the degrees of freedom of the ranging sensor.

Integrating a Part-Positioning Table. Addition of this hardware would allow remote sections of the seam to be brought within the robot's envelope and thus complex seams can be easily tracked. In addition, a coordinated motion of the robot arm and the part-positioning table would prevent high accelerations of the robot arm joints while tracking seams with a small radius of curvature.

From the welding process's perspective, a positioning table can allow changing the seam's orientation in a controlled fashion to keep the weld puddle horizontal for producing quality welds. Fernandez and Cook (Fernandez and Cook 1988) have presented a solution strategy for this problem when a CAD-database of the seam is available. In their approach, the combined hardware of the robot arm and the part-positioning table is treated as a single mechanism with eight degrees of freedom. For non-preprogrammed systems, the hybrid model of the seam geometry in the Base coordinate frame can replace the CAD-database in the above approach.

Additional Degrees of Freedom to the Range Sensor. The seam tracking strategy discussed so far allowed the range sensor a single degree of freedom, i.e., rotation about the torch axis. This limited movement of the sensor may restrict its ability to properly view the seam. In extreme cases, the sensor can lose track of the seam completely. To overcome this problem, future enhancements to the system could introduce more degrees of freedom for the ranging sensor by (i) adding more joints between the endeffector holding the torch-sensor assembly and the range sensor, or (ii) by forcing a lead/lag angle (within acceptable limits) on the torch. The second approach would, of course, orient the torch at an angle other than the normal to the seam but the view-vector can now be more favorable to scanning, as illustrated in Figure 9.1. Control schemes for implementing the above two approaches would be based on predicting the location of the next scan along the adaptive model of the seam. The first approach of introducing additional joints is apparently more promising but this requires integration of elaborate hardware and controller. On the other hand, the second approach of forcing a lead/lag angle on the torch is easier to implement with no additional hardware although it may provide rather modest improvements due to the limited range of acceptable lead/lag angle on the torch.

198

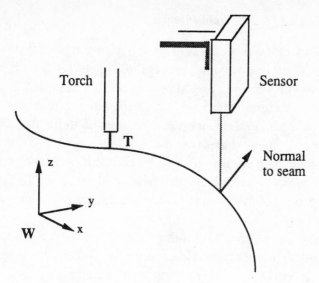

(a) Situation depicting an unfavorable view vector.

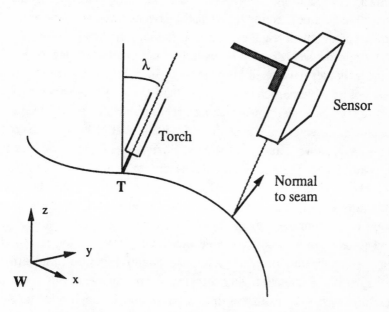

(b) Situation after forcing lead/lag angle (λ) on the torch.

Figure 9.1 Advantage of forcing a lead/lag angle on the torch

9.2.2 Integrating an Online Weld Planner

Adding flexibility to the seam tracker's operation invariably translates into realtime control of the welding process parameters. For instance, simultaneous welding and tracking seams with small radius of curvature requires reduced welding speed with corresponding adjustment to the wirefeed rate. The wirefeed rate must also be adjusted to fill the joint with correct volume of metal based on the joint geometry. Tack welds encountered along the seam may have to be burned through by increasing the arc voltage, or the wirefeed rate may have to be reduced. To facilitate such realtime control of the welding process parameters, an online welding planner is necessary.

Researchers in the area of welding automation have addressed the problem of welding process control in great detail. Welding has progressed much from the early manual welding with simple slag systems to semiautomatic welding with higher heat inputs, refined slag systems, and optimized composition. However, there are many indications that the traditional techniques used in the industry have matured and new areas such as intelligent automation and computer technology are being explored.

The approaches for adaptive welding automation are manifold. The intelligent welder developed by Automatix Inc., Massachusetts, USA uses simple lookup tables which recommend the proper welding parameters such as arc voltage, wirefeed rate, tracking speed, torch offsets and orientation, based on the measured joint geometry (Agapakis et al. 1990b). The creation of the lookup table initially requires extensive experimentation and input from welding experts. Other researchers have developed highly sophisticated weld planners based on artificial intelligence concepts and blackboards integrating large scale databases to decide upon the control strategy to be employed in the hybrid weld process control architecture. Such a system can incorporate a wide variety of control strategies including, model-based control, rule-based control, adaptive control, and artificial neural systems. Although highly complex, the advantage of this system lies in dealing with data from multiple sensors and in environments where uncertainty exists in both, the models and the sensor data.

Welding process control can also benefit from the use of multiple sensors, sensing not only the seam geometry but also the welding conditions and completed welds. This approach would provide a closed loop approach to weld quality control by (i) using the weld planner to set the weld parameters based on the high-level controller's model of the seam environment, then (ii) using the sensed weld condition in the innerloop of the feedback control, and (iii) also inspecting the area immediately behind the torch to compute the welding and tracking parameters in the outerloop of the controller.

9.3 ERROR ANALYSIS

The various sources contributing to the tracking error were discussed in Chapter 7. These errors are generated in different coordinate frames and an upper bound on the individual errors can be established through deterministic or stochastic approaches. However, the maximum upperbound of these individual errors cannot be directly used to compute the cumulative tracking error, which is strictly in the Base coordinate frame. Furthermore, the individual errors may contribute in a multiplicative manner, or they may be independent of each other and effectively sum up or cancel each other. It has already been shown in Chapter 7 that the tracking error does not propagate from one cycle to another. A detailed error analysis would help in establishing an optimum bound on the tracking error and further analysis could suggest possible ways to reduce these errors.

9.4 SUMMARY

The seam tracking capabilities of the seam tracker can be further enhanced by adding more degrees of freedom for positioning both, the torch and the sensor over the seam. Although the tracking error is shown to be uniformly bounded during the entire seam tracking operation, a detailed analysis of the tracking error during any cycle can be used to understand and further improve the system performance. In the context of robotic welding, an online welding planner can be integrated with the high-level controller for realtime control of the welding process parameters.

REFERENCES

Agapakis, J.E. (1990a) Approaches for recognition and interpretation of workpiece surface features using structured lighting. International Journal of Robotics Research, Vol. 9 No. 5, October 1990, pp. 3–16.

Agapakis, J.E., Katz, J.M., Friedman, J.M., and Epstein, G.N. (1990b) Vision-aided robotic welding: An approach and a flexible implementation. International Journal of Robotics Research, Vol. 9 No. 5, October 1990, pp. 17–34.

Albano, A. (1974) Representation of digital contours in terms of conic-arcs and straight-line segments. Computer Graphics and Image Processing, pp. 23–33.

Arata, Y. and Inoue, K. (1975) Automatic control of arc welding (Report 5). Transactions JWRI 1975, Vol.4, No.2, pp. 101s–104s.

Begin, G., Boillot, J.P., Michel, C., and Teubel, G. (1985) Third generation adaptive robotic arc welding unit in Proc. Intl. Inst. Welding, Annual Conf., Strasbourg, France, September 1985, pp. 401–411.

Bjorkelund, M. (1987) A True Seam Tracker for Arc Welding. Robotic Welding, ed. J.D. Lane, IFS Publications, UK, pp. 167–180.

Boillot, J.P. (1985) Adaptive welding by fiber-optic thermographic sensing: An analysis of thermal and instrumental considerations. Welding Journal, Vol. 64, pp. 209s.

Bolles, R.C., and Fischler, M.A. (1981) A RANSAC-based approach to model fitting and its application to finding cylinders in range data. In Proc. 7th Intl. Joint Conference on Artificial Intelligence, Menlo Park, CA. pp. 637–643.

Bollinger, J.G. and Harrison, H.L. (1981) Automated welding using spatial seam tracings. Welding Journal, Vol. 50, November 1981, pp. 787–792.

Brown, K.W. (1975) A technical survey of seam tracking methods in welding. Welding Institute Report 3359/1-3/1973.

Chang, T.C., Wysk, R.A. and Wang, H.P. (1991) Computer-aided Manufacturing. Prentice-Hall, Englewood Cliffs, NJ, 1991.

Clocksin, W.F., Bromley, J.S.E., Davey, P.G., Vidler, A.R., Morgan, C.G., (1985) An implementation of model-based visual feedback for robot arc welding of thin sheet steel. International Journal of Robotics Research, Vol. 4, No. 1, Spring 1985, pp. 12–26.

Cook, G.E. (1980) Feedback control of process variables in arc welding. Proceedings 1980 Joint Automatic Control Conference, New York, Vol. 2, August 1980, pp. 9.

Cook, G.E. (1981) Feedback and adaptive control in automated arc welding systems. Metal Construction, Vol. 13, No. 9, pp. 551–555, September 1981.

Cook, G.E. (1983) Robotic arc welding: Research in sensory feedback control in IEEE Trans. on Indl. Electronics, Vol. IE-30, No.3, August 1983, pp. 252–268.

Cook, G.E., Yizhang, L., and Shepard, M.E. (1986) Computer-based analysis of arc welding signals for tracking and process control. IEEE Trans. on Ind. Electronics, Vol.34, 1986, pp. 1512–1518.

Corby Jr, N.R. (1984) Machine Vision Algorithms for vision guided robotic welding. Proceedings of the 4th International Conf. on robot vision and sensory controls (ROVISEC4), London, pp. 137–147.

Davies, B.R., Red, W.E., and Lawson, J.S. (1990) The local calibration method for robot inaccuracy compensation. Journal of Robotic Systems, John Wiley & Sons, Inc. pp. 833–864.

Day, C.P. (1988) Robot accuracy issues and methods of improvement. Robotics Today (RI/SME's Quarterly on Robotics Technology), Vol. 1 No. 1, Spring 1988. pp. 1–9.

Denavit, J. Hartenburg, R.S. (1955) A kinematic notation for lower-pair mechanisms based on matrices. ASME Journal of Applied Mechanics, June 1955, pp. 215–221.

Drews, P. and King, F.J. (1975) Automatic seam tracking in arc welding by an optical method presented at Second Int. JWS Symposium, August 1975.

Estochen, E.L., Neuman, C.P., and Prinz, F.B. (1984) Application of acoustic sensors to robotic seam tracking. IEEE Trans. Ind. Electronics, Vol. IE-31, No. 3, August 1984.

Fenn, R. and Stroud, R.R. (1986) Development of an ultrasonically sensed penetration controller and seam tracking system for welding robots, in UK Robotics Res. 1984, IMECHE Conf. Pub. London: Mechanical Engineering Publications, Ltd., 1986, pp. 105–108.

Fernandez, K.R., and Cook G.E. (1988) A generalized method for automatic downhand and wirefeed control of a welding robot and positioner. NASA Technical Paper 2807, February 1988.

Frober, C.E. (1969) Introduction to Numerical Analysis, 2nd edition, Addison-Wesley, Reading, Massachusetts, 1969.

Gennery, D.B. (1982) Tracking known three-dimensional objects. Proceedings of AAAI-82, pp. 13–17.

Gonzalez, R.C. and Winz, P. (1987) Digital Image Processing. Addison Wesley, Reading, Massachussetts.

Goodwin, F.E. (1985) Coherent Laser Radar 3-D Vision Sensor SME Technical paper MS85-1005, Society of Manufacturing Engineers, Dearborn, MI.

Horaud, R. and Bolles, R.C. (1983) Two dimensional curve segmentation. Twelfth report of SRI International on machine intelligence applied to industrial automation, Menlo Park, CA.

Huber, C. (1987) Sensor-based tracking of large quadrangular weld seam paths. Robotic Welding, Edited by J.D. Lane, IFS(Publications) Ltd. UK, 1987, pp. 141–156.

Jones, S.B., Holder, S.J., and Weston, J. (1981) Mechanical approaches to seam tracking for arc welding. Welding Inst. Res. Report 167/1981.

Juds Scott M. (1988) Photoelectric Sensors and Controls, Marcel Dekker, Inc., New York, 1988, pp. 29–32.

Justice, J.F. (1983) Sensors for robotic arc welding in Proc. AWS Conf. Automation and Robotics For Welding, February 1983, pp. 203–210.

Kawahara, M. (1983) Tracking control system using image sensor for arc welding. Automatica, Vol 19 No 4 pp. 357–363.

Kreamer, W., Kohn, M., and Finlay, J. (1986) Vision-guided robotic turret welding system. Proceedings of the Robots 10 Conference, Chicago, Illinois (Sponsored by Robotics International of SME, Dearborn, Michigan), April 20-24, 1986, pp. 13–26.

Lane, J.D. (1987) Editor Robotic welding–International trends in manufacturing technology, IFS Publications, UK and Springer Verlag, Berlin, 1987.

Lau, K., Hocken, R. and Haynes, L. (1985) Robot performance measurements using automatic laser techniques. NBS-Navy NAV/CIM workshop on robot standards, Detroit, MI, June 6,7 1985.

Levine, M.D. (1985) Vision in man and machine. McGraw Hill, New York.

Liao, Y.Z. (1981) A two-stage method of fitting conic arcs and straight-line segments to digitized contours. In Proc. Pattern Recognition and Image Processing Conf. New York (IEEE) pp. 224-229.

Linden, G., and Lindskog, G. (1980) A control system using optical sensing for metal-inert gas arc welding. Proceedings TWI Conf., November 1980.

Morrison, N. (1969) Introduction to sequential smoothing and prediction, McGraw Hill Publishers, New York.

Moss, J.A. (1986) Attributes and limitations of a laser-based vision system for robotic arc welding. Proceedings of the Robots 10 Conference, Chicago, Illinois, (Sponsored by Robotics International of SME, Dearborn, Michigan), April 20–24, 1986, pp. 11–22.

Nayak, N.R. (1989) An integrated control system for intelligent seam tracking in robotic welding, Doctoral Dissertation, Department of Mechanical Engineering, The Pennsylvania State University, 1989.

Nayak, N.R., Vavreck, A., Kuhlman, D. and Salvi, M. (1990) Robotic inspection and adaptive flame-cutting of large parts for welding joint preparation, International Journal of Robotics and Automation (IASTED), Vol. 5 No. 2, pp. 81–91.

Niepold, R. (1979) Observation of CO_2 dip transfer arc welding with a television sensor for remote control and automation of the welding process. Welding Institute Translation, (Original pub. Deutscher Verband for Schweisstechnik und Schneiden, 1979, pp. 38–43)

Niepold, R. and Brummer, F. (1987) PASS–A Visual Sensor for Seam Tracking and On-Line Process Parameter Control in Arc Welding Applications. Robotic Welding, ed. J.D. Lane, IFS Publications, UK, pp. 129–140.

Nitzan, D., Bolles, R., Cain, R., Hannah, M., Herson, J., Horaud, P., Kremers, J., Myers, J., Park, W., Reifel, S., and Smith, R. (1983) Machine intelligence research applied to industrial automation, Technical Report No. 12 SRI International, Menlo Park, California.

Nitzan, D., Bolles, R., Kremers, J., and Mulgaonkar, P. (1987) 3-D Vision for Robot Applications. In NATO ASI Series F33 on Machine Intelligence and Knowledge Engineering for Robotic Applications, (Ed) AKC Wong and A. Pugh, Springer-Verlag Berlin Heidelberg, pp. 21–81.

Oomen G.L. and Verbeek WJPA (1987) A Realtime Optical Profile Sensor for Arc Welding Robots. Robotic Welding, ed. J.D. Lane, IFS Publications, UK, pp. 117–128.

O'Rourke, J.T. (1987) A Case for CIM, Proceedings of the National Science Foundation Workshop on Computer Networking for Manufacturing Systems, ed. A. Ray, Pennsylvania State University, University Park, PA, November 1987.

Paul, R.P. (1977) The mathematics of computer-controlled manipulation. The 1977 Joint Automatic Control Conference, July 1977, pp. 124–131.

Paul, R.P. (1981) Robot manipulators: Mathematics, programming, and control, MIT Press, Cambridge, Massachusetts, 1981.

Pavlidis, T. (1982a) Algorithms for graphics and image processing. Computer Science Press, Rockville, MD.

Pavlidis, T. (1982b) Curve fitting as a pattern recognition problem. Proceedings of the 6th International Conf. on Pattern Recognition, IEEE New York, pp. 853–859.

Pavone, J.G. (1983) Univision II, a system for arc welding robots in Proc. AWS Conf. Automation and Robotics for Welding, February 1983, pp. 91–104

Peterson, J.L. (1981) Petrinet Theory and the Modeling of Systems, Prentice Hall Inc., Englewood Cliffs, New Jersey, 1981.

Pieper, D.L. (1968) The kinematics of manipulators under computer control. Stanford Artificial Intelligence Laboratory, Stanford University, AIM 72, 1968.

Porsander, T. and Sthen, T. (1987) An adaptive torch positioner system. Robotic Welding, ed. J.D. Lane, IFS Publications, UK, pp. 157–166.

Raina, Y. (1988) Double electrode seam tracking system using through-the-arc technique. Masters Thesis, The Pennsylvania State University.

Ramer, U. (1972) An iterative procedure for polygonal approximation of plane curves. Computer graphics and image processing, pp. 244–256.

Richardson, R.W. (1982) Seam tracking sensors–Improving all the time. Welding Design Fabrication, pp. 77–82, September 1982.

Rosenfeld, A. and Kak, A.C. (1982) Digital Picture Processing. Academic Press, New York.

Ross, B. (1984) Machines that can see: Here comes a new generation. Business Week, January 1984, pp. 118.

207

Smati, Z., Yapp, D., Smith, C.J. (1984) Laser guidance system for robots in Proceedings of the 4th Intl Conf. on Robot Vision and Sensory Controls (ROVISEC4), London, October 1984, pp. 91–101.

Tomizuka, M., Dornfeld, D., and Porcell, M. (1980) Application of microcomputers to automatic weld quality control. J. Dyn. Sys., Meas.,Contr., Vol.102, No.2, pp. 62–68.

Tsuji, S. (1983) Wiresight: Robot vision for determining three-dimensional geometry of flexible wires in 1983 Int. Conf. on Advanced Robotics, Tokyo, Japan, pp. 133–138.

Umeagukwu, C. Maqueira, B., and Lambert, R. (1989) Robotic acoustic seam tracking: system development and application. IEEE Trans. on Industrial Electronics, Vol. 36, No. 3, August 1989, pp. 338–348.

Vanderbrug, J.G., Albus, J.S., and Barkmeyer, E. (1979) A vision system for realtime robot control. Robotics Today, Winter 1979–1980, Vol. 1, No. 6, pp. 20.

Verdon, D.C., Langley, D., and Moscardi, D.H. (1980) Adaptive welding control using video signal processing in Dev in Mech. Automated and Robotic Welding, presented at TWI Seminar (London), November 1980.

Waller, D.N., Foster, C.J., and Wagner, R. (1990) Realtime imaging for arc welding, Int. Journal of Computer Integrated Manufacturing, 1990, Vol. 3, Nos. 3 and 4, pp. 249–260.

Weeg, G.P. and Reed, G.B. (1966) Introduction to numerical analysis, Blaisdell Publishing Co., Waltham, Massachusetts, 1966.

Westby, A.O. (1977) An adaptive controlled welding machine. SINTEF Rep. STF 16 A77028, January 1977.

OTHER REFERENCES

Greenleaf Comm Library User's Manual (version 2.0). Greenleaf Software Inc., Carrolton, Texas, USA.

IBM PC/AT VME Adapter Documentation, BiT3 Computer Corporation, Minneapolis, Minnesota, USA.

IBM Disk Operating System Version 3.10 Reference, International Business Machines Corporation, Boca Raton, Florida, USA.

Installation Guide to Miller Electric Deltaweld 450, Miller Electric Mfg. Co., Appleton, Wisconsin, USA.

Installation Guide to Miller Electric S54A, Miller Electric Mfg. Co., Appleton, Wisconsin, USA.

Laser Profile Gauge Overview Document, Chesapeake Laser Systems, Inc., Lanham, Maryland, USA.

Lattice C Compiler for 8086/8088 Series Microprocessor (Document revision 2.15A), Lattice Inc., Glen Ellyn, Illinois, USA.

Owner's Manual and Guide to Operations–PC's Limited 286[12], PC's Limited, Texas, USA.

Programming Manual User's Guide to VAL II (version 1.1) Unimation Inc., Danbury, Connecticut, USA.

VMEbus Specification Manual (revision c.1) compiled and edited by Micrology pbt, Arizona, for the VMEbus International Trade Association.

INDEX

                              211

212

ESPRIT project on realtime imaging for arc welding 195–6

Gas-metal arc welding (GMAW) 11, 23
Gas-tungsten arc (GTA) welding 23
Geometric model of seam. *see under* Seam geometry
GMAW 11, 23
GTA 23

H

Heat resistant brushes 23
High-level controller 131
Horn concentrator (ultrasonic ranging) 184

I

Image processing
  AND operation in 33
  gaussian filter 32
  histogram equalization 32
  image segmentation 44
  intensity thresholding 33
  logarithmic transformation 32
  median filter 32
  of laser stripe images
    least mean square slope detection filter 50
  of vee-joint range images 88–102
  of vee-joint range images, algorithms 90
  spatial filtering 33
Image space 31, 44, 87
Indirect range measurement
  from shadows 20
  suitability to realtime applications 19
  through focussing 20
  using known object geometry 20
  using moiré fringe patterns 21
Indirect range measurement 19–21
Initial path tracking 147
Intelligent realtime control (IRTC) 27
Intensity image
  error
    in image coordinates 31, 34

219

S

U

Ultrasonic ranging 181–3
Unimation, Inc. 137

V

VAL robot control language 137, 145
Vee-grooved weld joint . *see under* Welding seam
VME bus 132, 136

W

Weld path generation. *see* Low-level control
Welding seam
    minimum radius of curvature specifications
        based on allowable tracking error 163–4
        based on range sensor's field of view 164–6
        based on torch-sensor assembly design 166–8
    types
        fillet joint 41, 48
        split joint 41, 42, 43
        vee-joint 44, 45, 46, 47, 56, 57, 84, 88, 89, 93, 99, 102
World space 31

Heterick Memorial Library
Ohio Northern University

DUE	RETURNED	DUE	RETURNED
1. OCT 11 95	OCT 18	13.	
2. FEB 15 1999	FEB 3 1999	14.	
3. FEB 05 20	APR 29 20 15.		
4. MAY 9 2002		16.	
5.		17.	
6.		18.	
7.		19.	
8.		20.	
9.		21.	
10.		22.	
11.		23.	
12.		24.	

WITHDRAWN FROM
OHIO NORTHERN
UNIVERSITY LIBRARY